Janus
Physical Science

GLOBE FEARON EDUCATIONAL PUBLISHER
A Division of Simon & Schuster
Upper Saddle River, New Jersey

CONTRIBUTORS

Cary I. Sneider, Ph.D.,
Lawrence Hall of Science,
University of California, Berkeley
Elaine Katz
Janis Fisher Chan

Henri Picciotto,
The Urban School,
San Francisco, California
Richard S. Kolbert
Susan Echaore-Yoon

Budd Wentz

CONSULTANTS

Gilbert Yee, Fremont, California
Donald W. Pettibone, Ph.D., Diasonics, Inc., Milpitas, California
Cary Sneider, Ph.D., Lawrence Hall of Science, University of California, Berkeley

REVIEWER

Anna Marie Villalobos, Mount Pleasant Hill High School, San Jose, California

ARTISTS

Ellen Beier, Nancy Kirk, Winfield Coleman, Marilyn Hill, Don Bauders, Margaret Sanfilippo

PHOTO CREDITS

cover: © J. Zimmerman, FPG, 3: Dan Suzio, Photo Researchers, 18: The Field Museum, Chicago, IL Neg. # Geo 85830 Photo by John Weinstein, 49: Grant Heilman, 95: Mark Burnett, Photo Researchers, 141: Photo Researchers

ISBN 0-835-91390-2

Printed in the United States of America
1 2 3 4 5 6 7 8 9 10   99 98 97 96 95

GLOBE FEARON EDUCATIONAL PUBLISHER
A Division of Simon & Schuster
Upper Saddle River, New Jersey

Contents

Introduction

In this book, you will learn about some of the things that scientists study. You will learn some of the things that scientists have discovered. You will explore and discover facts the same way scientists do, by experimenting, observing, and recording. And you will learn scientific information about the world that you can use right away.

Forces in Our World

Physical science is the study of the forces in our world. You need those forces to stay alive. And you can use the forces to make your life better.

Before people understood the forces in the world, their life was much harder than our life is today. Before people knew how to make fire, they weren't able to cook their food or heat their homes. Before machines were invented, work was much harder.

Physical scientists study the forces in our world. They have learned some of the ways those forces work. And they have discovered ways we can use those forces to make our life better.

Scientists have discovered how energy, such as electricity, works. You use energy to light your house or to travel from one place to another.

Scientists have discovered how sounds are made and how sounds travel. You can listen to music or talk on the telephone because of what scientists have discovered about sound.

And scientists have invented machines for you to use. You can move heavy things or cut the grass because of what scientists have learned about how machines work.

In this book, you'll learn about different forces in our world. As you study those forces, you'll use the same methods that scientists use: You will observe how the forces work and record what you learn. And you will learn some of the ways you can control those forces to make your life better.

ENERGY

What kind of force is energy? What kinds of energy do we use? Where does that energy come from? How do we use energy? How can we save energy so that it doesn't run out? In this section, you'll learn many facts about energy. And you'll learn about the important part energy plays in our lives.

Contents

Introduction

Picture this:

It's late afternoon. It is time to go home, and you're waiting for a bus.

You look around. Cars and buses fill the street. Motorcycles roar by. An airplane flies overhead.

The sun sets, and you feel cold. You put on your sweater. Street lights come on. Lights in buildings come on too.

Your bus pulls up, and you get on. Soon you'll be home. You'll turn on the heater and warm your house. You'll cook your dinner, then watch TV. At bedtime, you'll set the alarm on your electric clock.

All those things take energy. Energy moves the cars, buses, motorcycles, and airplane. Energy gives you light and heat. Energy runs your TV and clock.

Energy is all around us. We use it to make our lives easier and more comfortable. We use it to make things and to do work. We use it to keep ourselves alive. Can you imagine living without any energy at all?

This section is about the energy you use. You'll find out where that energy comes from. You'll learn how it is used. You'll see how important energy is in your life. And you'll also learn ways to use energy wisely.

Unit 1

What Is Energy?

You use energy in many different ways. You use it to cook your food. You use it to get to school or work. And you use it to light up a dark place. In fact, you use a *lot* of energy every day.

Even though you use energy all the time, you probably don't think about it much. But without energy, you could not live. Energy is all around you. You use many kinds of energy.

- What are some kinds of energy?
- How do you use energy?
- Where do you get energy?

You'll learn the answers in this unit.

Before You Start

You'll be using the science words below. Find out what they mean. Look them up in the Glossary that's at the back of this book. On a separate piece of paper, write what the words mean.

1. **motion**
2. **stored**

Stuck in the Woods!

Picture this:

You're driving down a road. Your car is moving along. It's in *motion*.

The sky starts to get dark. Night is coming. You can't see the road very well, so you turn on the car's headlights. They *light* the road.

Now it starts to snow. You're getting cold. You turn on the car's heater. It sends *heat* into the car. Soon you are nice and warm.

You're now using three kinds of energy. What do you think they are?

Right! Motion, light, and heat are the three kinds of energy you're using.

All of a sudden, your car stops! Your lights go out, and your heater stops working. You can't get the car to start up again. You are cold, you have no light, and you are stuck in the woods. You have an energy problem! What would you do?

That's an example of what could happen when you can't get motion, light, and heat energy. They are three kinds of energy you use every day. Without them, your life could be very hard. What would your life be like if you didn't have motion, light, and heat energy?

What Gives You Energy?

You learned that heat, light, and motion are kinds of energy. What are some things that give you those kinds of energy?

Heat Energy

Name something that gives off heat to warm you.

Name something that gives you heat to cook with.

Light Energy

Name something that gives you light indoors.

Name something that gives you light outdoors.

Name something you can carry that gives you light.

Motion Energy

Name something that moves you from home to school or work.

Name something that moves you on a river, lake, or ocean.

Name something that helps people move heavy things.

Changing Energy

Think of a piece of paper. Does it have motion, heat, or light energy?

It doesn't have any of those kinds of energy. But it can! That's because paper has **potential energy**. Potential energy is a kind of energy that is stored inside of something.

Scientists say that energy cannot be made. It can only be *changed* from one kind of energy into another kind. How can the potential energy in paper be changed into light energy?

Right! When you burn paper, the potential energy inside it is changed into light energy. It is also changed into heat energy.

The pictures on this page show some things that had potential energy. What kinds of energy did that potential energy change into? Answer the questions. Then check your answers. (The right answers are upside down.)

1. Wood has potential energy. When you burn wood, the potential energy is changed into what two kinds of energy?

   a. heat   b. light   c. motion

2. Gasoline has potential energy. When you drive a car, the engine burns gasoline. The car moves. The potential energy in gasoline is changed into what kind of energy?

   a. heat   b. light   c. motion

3. A battery has potential energy. When you use it in a flashlight, that potential energy is changed. What kind of energy is it changed into?

   a. heat   b. light   c. motion

Answers

1. We burn wood to get heat and light energy.
2. We get motion and heat energy when we burn gasoline.
3. We get light and heat energy when we use a battery in a flashlight.

Review

Use what you learned in this unit. Match the words in the list below with their meanings. The page number after each meaning tells where you can find the word.

potential   motion
heat   light

1. A kind of energy that can move things (page 7)
2. A kind of energy that helps you see things (page 7)
3. Stored energy that is waiting to be used (page 9)
4. A kind of energy that can cook food (page 7)

Check These Out

1. Make a Science Notebook for Energy. Keep a list of your glossary words and their meanings in your notebook. Also keep a record of the experiments you do. You can also put anything else you learn about energy in your Science Notebook.

2. Find out more about how energy works in a car. Ask a mechanic to talk to your class. Ask the mechanic these questions: How does the car's heater work? Where do the headlights get their energy? How does the engine work?

3. Look through some magazines. Find pictures that show energy being used. Cut them out and put them in these three groups: heat, light, and motion. Make posters of the three groups.

4. Food gives people energy to move. This energy is measured in *calories*. Find out how many calories certain foods have. Get a food calorie chart from a library. Or look at the labels on cans and other food packages. Those labels often list the number of calories in the food. Bring those labels to class.

5. As you work through this section, you may want to find out more about energy. You can find out more by looking in a dictionary or an encyclopedia, or by getting books about energy from a library. You can also talk to an expert, such as a physics teacher, someone at a utility company, or someone at an ecology center.

Here are some things you may want to find out:
- What is the science of thermodynamics? How do engineers use this science?
- What does *conservation of energy* mean? Give some examples.
- What is horsepower?
- What kinds of energy did cave people use? What kinds of energy do people use today?

Unit 2

Burning Fuels

Fire! It is the most important energy discovery of all time. Fire gives us heat and light energy. It also gives us motion energy.

But in order to have a fire, something must burn. We burn many different things every day. They are our most important sources of energy.

- What do people burn for energy?
- How do people use the energy from burning things?

You'll learn the answers in this unit.

Before You Start

You'll be using the science words below. Find out what they mean. Look them up in the Glossary. On a separate piece of paper, write what the words mean.

1. **engine**
2. **fuel**

Energy from Fuel

We call things that we burn *fuel*. A good fuel has a lot of potential energy stored in it. When we set that fuel on fire, the fire changes its potential energy into heat, light, or motion.

The pictures show some fuels being burned. They are fuels people use every day. Look at the pictures and answer the questions. Then check your answers. (The right answers are upside down.)

1. The people in the top picture are having a barbecue. What fuel are they burning?

    What energy is that fuel giving?

2. The power went out in the middle picture. The lights don't work. What fuel are the people burning?

    What energy is that fuel giving?

3. The woman in the bottom picture is riding a motorcycle. The motorcycle has an engine. The engine works by burning fuel. What fuel is the engine burning?

    The engine changes some of the potential energy of the fuel into another kind of energy. What kind is it?

Answers

1. People burn charcoal to get *heat* energy.
2. People burn candle wax to get *light* energy.
3. The engine burns gasoline and gets *motion* energy.

The Right Fuel for the Job

People use different fuels for different jobs. Some fuels are used for light. Some are used for heat. And some are used for motion.

The pictures on this page show six fuels that are being used for different reasons.

1. Which two fuels are being used for light?
2. Which two fuels are being used for motion?
3. Which two fuels are being used for heat?

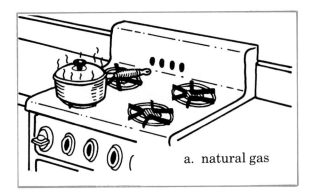

a. natural gas

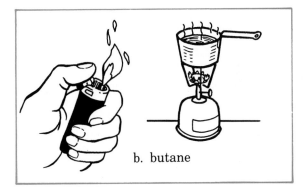

b. butane

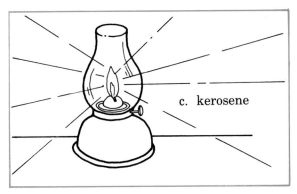

c. kerosene

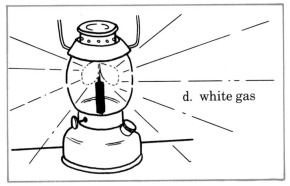

d. white gas

e. diesel fuel

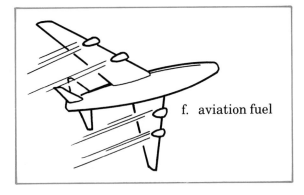

f. aviation fuel

Experiment 1

Which fuel is better for cooking?
Which fuel is better for light?

Materials (What you need)

Four short, thick candles

One can of Sterno

Water

Two small foil tart pans

Two whole foil loaf pans

One stapler

Scissors

One box of long matches

One felt pen

One foil loaf pan cut in half

Procedure (What you do)

1. Draw a hole on the bottom of each half pan. The hole should be a little bigger than the bottom of a tart pan. Then cut out the hole.

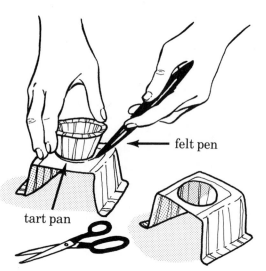

felt pen

tart pan

2. Make two "stoves" like this: Staple a half pan to each whole pan.

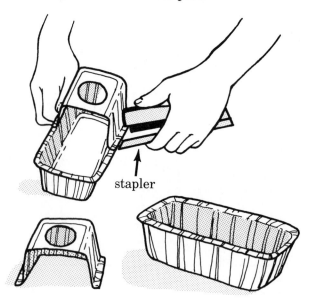

stapler

3. Fit a tart pan into the hole on each stove. Fill the pans about half full with water.

4. Put the candles under one tart pan. Take the lid off the Sterno and put the can under the other tart pan. Light the candles and the Sterno. Wait until the water boils.

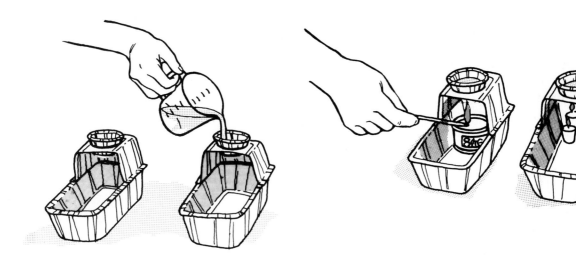

Observations (What you see)

1. Which stove boils water first: the one with Sterno or the one with candles?
2. Which stove gives off more light?
3. Which stove gives off more heat? How do you know?

Conclusions (What you learn)

1. Which fuel is better for cooking? Why?
2. Which fuel is better for light? Why?

Important!

Put out the flames.

1. Slide the Sterno lid onto the can. The flame will go out right away. <u>Do not blow out the Sterno flame!</u>
2. Blow out the candles.

Review

Show what you learned in this unit. Finish these sentences. The words you'll need are listed below.

burned   candle wax   fuels
engines   natural gas   potential
diesel fuel

1. Wood, gasoline, and butane are _____.
2. Fuels have _____ energy stored inside of them.
3. A fuel gives energy when it is _____.
4. Fuels burn in the _____ of cars.
5. Charcoal and _____ _____ are burned for heat energy.
6. Gasoline and _____ _____ are burned for motion energy.
7. White gas and _____ _____ are burned for light energy.

Check These Out

1. Try to hard-boil an egg on one of the stoves you made for the experiment on pages 14–15. Use the fuel that makes more heat. Cook the egg in boiling water for about ten minutes. Turn it over with a spoon, and cook it five more minutes. Then eat your egg.

2. Which is better fuel for a barbecue: wood or charcoal? Find out—*outdoors*. You will need two barbecue grills and two pieces of the same kind of meat. The pieces should be the same size. Start a wood fire in one grill and a charcoal fire in the other. Cook a piece of meat on each fire. Compare the results.

3. What are regular, premium, and unleaded gasolines? Ask someone who works at a gas station.

4. Invite someone from the fire department to talk to your class. Ask that person how to store fuels safely to prevent fires.

5. People from different countries have legends and myths about fire. Do you know one? Tell it to your class.

6. What are good safety rules about using fire? Make a poster.

7. Here are more things you may want to find out.
   - What is combustion? What is spontaneous combustion?
   - How does an internal combustion engine work?
   - What happens to fuel when it burns? What is oxidation?
   - How do matches work?
   - What's in Sterno? Why does it burn?

Unit 3

Fuels from Under the Ground

You depend on cars, buses, and home furnaces every day. Those things work because they burn fuels that come from under the ground.

Your clothes and many of the things you use are made by machines in factories. Those machines work because of fuels that come from under the ground.

Our way of living today depends on burning huge amounts of fuel. Almost all of that fuel comes from under the ground.

- What fuels come from under the ground?
- How do we use those fuels?
- How do we get those fuels?

You'll learn the answers in this unit.

Before You Start

You'll be using the science words below. Find out what they mean. Look them up in the Glossary. On a separate piece of paper, write what the words mean.

1. **fossil**

2. **mine**

3. **products**

Fossil Fuels

Imagine Earth millions of years ago, long before there were people. Many of the plants and animals were different from the kinds we see today. Strange-looking fish filled the oceans. Giant dinosaurs walked through forests. Those forests covered much of the land.

All those plants and animals died. Many were buried under mud and sand. As years went by, layers of dirt and rocks covered them. For millions of years, tons of dirt and rock pressed down. What happened to those dead plants and animals?

Right! They began to rot. And then they changed. They slowly turned into materials that could be burned. Those materials are called **fossil fuels**. Why are they called *fossil* fuels?

Fossils are the remains of plants and animals that lived millions of years ago. So fossil fuels are the remains of ancient living things that have turned into fuels.

A fossil fuel can be a gas, a liquid, or a solid.

Gas fossil fuel is called **natural gas**. Natural gas is used for cooking and for heating homes.

Liquid fossil fuel is called **crude oil**. Crude oil is made into gasoline, butane, kerosene, aviation fuel, diesel fuel, and other products.

Coal is one kind of solid fossil fuel. It is used in factories to make steel, cement, paper, and other things we use every day.

How would your life be different if there were no fossil fuels? (Think about how you get around town, cook your food, or heat your house.)

Digging for Coal

Burning a rock? You may find that surprising, but it happens every day. Coal is a black rock that makes a lot of heat when it is burned. Coal is also a fossil fuel.

Once coal was used mainly to run trains and to heat houses. Today it is burned mostly in factories to make steel, cement, paper, and other things.

Coal is found in the ground, like other fossil fuels. People get coal out of the ground by digging huge mines deep into the earth. People who work in mines are called *miners*.

Look at the picture. It shows a *shaft mine*. (A shaft is a deep hole.) The shaft is dug into the earth, down to the coal. How do the miners get down the shaft?

Right! The miners use an elevator to get down to the coal. There they dig out the coal. As they dig farther and farther into the coal, they make a tunnel. The tunnel may be several miles long. What do the miners use to help them move the coal through the tunnel?

Yes! Miners use *shuttle cars* to move the coal to the shaft. There it is brought to the surface on the elevator. Then trucks, trains, and ships take it to the factories where it is used.

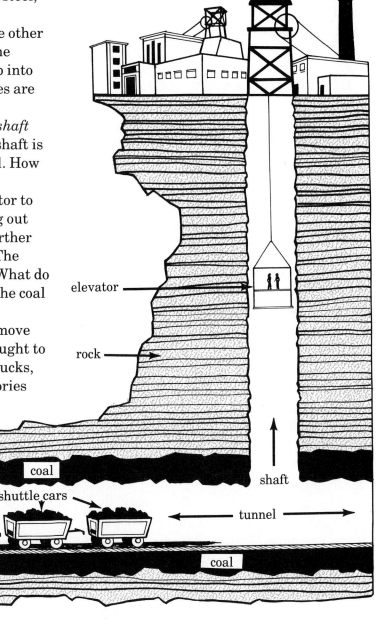

elevator

rock

coal

shuttle cars

coal

shaft

tunnel

rock

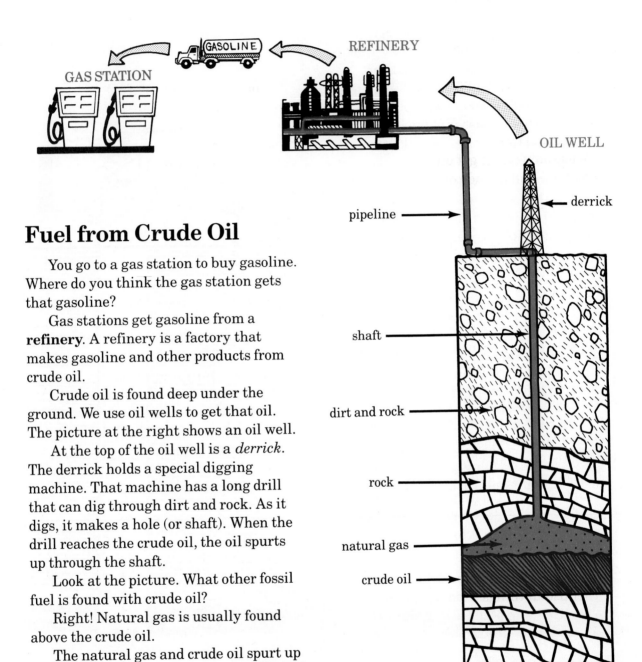

Fuel from Crude Oil

You go to a gas station to buy gasoline. Where do you think the gas station gets that gasoline?

Gas stations get gasoline from a **refinery**. A refinery is a factory that makes gasoline and other products from crude oil.

Crude oil is found deep under the ground. We use oil wells to get that oil. The picture at the right shows an oil well.

At the top of the oil well is a *derrick*. The derrick holds a special digging machine. That machine has a long drill that can dig through dirt and rock. As it digs, it makes a hole (or shaft). When the drill reaches the crude oil, the oil spurts up through the shaft.

Look at the picture. What other fossil fuel is found with crude oil?

Right! Natural gas is usually found above the crude oil.

The natural gas and crude oil spurt up the shaft. At the top of the ground, they go into a long pipe. What is that pipe?

Gas and crude oil from the shaft go into the *pipeline*. It takes the crude oil to a refinery. There some of the oil is turned into gasoline. How does the gasoline get to the gas station?

Review

Show what you learned in this unit. Finish the sentences. Match the words on the left with the correct words on the right.

1. Most of our energy
2. Some fossil fuels are
3. Fossil fuels are the remains of things
4. We get coal by
5. We use oil wells to get oil
6. Refineries turn crude oil

a. and natural gas out of the ground.
b. that lived millions of years ago.
c. comes from fossil fuels.
d. into gasoline and other products.
e. coal, oil, and natural gas.
f. mining it out of the ground.

Check These Out

1. Have you seen a movie or TV show about working in a coal mine? Tell the class about it. What are some problems that miners have?
2. The United States has a lot of coal under its surface. Find out where. Draw a map showing the states that have coal.
3. Find out the main places in the world where crude oil is found. Get a map of the world. Mark those places on the map with pins and stickers.
4. Is there a refinery, oil well, or coal mine in your area? Visit it with your class.
5. Crude oil is sometimes called *petroleum*. We make fuels and many other useful things from petroleum. What are those other useful things?
6. Many science museums have exhibits about fossil fuels. Visit one of those museums.
7. Here are more things you may want to find out:
   - Strip mining is another way of getting coal out of the ground. What is a strip mine like? How is it different from a shaft mine?
   - Oil is sometimes found under the ocean floor. How is that oil brought to the surface of the ocean?
   - How do refineries make gasoline from crude oil?
   - What other kinds of fuel are made from crude oil?
   - What happens to the natural gas that comes up through the shaft? How is it used? How does it get to a person's house?

Unit 4
Electricity

You learned that you can get energy by burning a fuel. But there's another way you can get energy. You can turn on a switch! For example, whenever you turn on a light switch, you are getting energy.

The energy you use is electrical energy. We call that energy *electricity.* You use it every day.

- How is electricity used?
- How is electricity changed?
- How is electricity made?

You'll learn the answers in this unit.

Before You Start

You'll be using the science words below. Find out what they mean. Look them up in the Glossary. On a separate piece of paper, write what the words mean.

1. **appliance**
2. **magnet**
3. **produce**

Electrical Energy

You turn on a switch to light a lamp. You turn on a switch to run a TV or an electric heater. Things such as lamps, TVs, and heaters are appliances. What do you think appliances do to electrical energy?

Right! They change electrical energy. They change it into heat, light, or motion energy.

When you turn on a light switch, electricity goes into a bulb. The bulb changes the electricity into light energy. Suppose you turn on a heater switch. What energy does the heater change the electricity into?

Suppose you turn on a washing machine switch. What energy does the machine change electricity into?

Now think about different appliances in your home, school, or work.

1. What appliances change electricity into *heat*?
2. What appliances change electricity into *light*?
3. What appliances change electricity into *motion*?

Electricity from a Battery

Turn on a switch—and you get electricity. But where does the electricity come from? It is *produced* by something.

One of the things that produces electricity is a battery. A battery has chemicals in it. It changes the energy of the chemicals into electricity.

You can see how it works. You'll need these materials:

- Two bell wires about 6 inches long
- One sharp knife
- One very small socket (with 2 posts)
- One flashlight bulb that fits the socket
- One flashlight battery (1.5 volts)
- One wire stripper or sharp knife

Caution: Be careful when you use a knife.

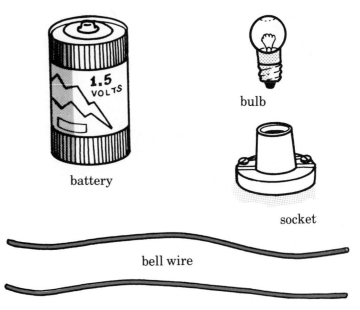

battery

bulb

socket

bell wire

1 Strip the ends of each wire. First, carefully cut the covering one inch from the end. (Don't cut through the wire.) Cut all around the wire.

Pull the covering off the wire. You should now have an inch of metal wire.

2 Get the socket. Connect one metal end of one wire to a post. Connect one metal end of the other wire to the other post.

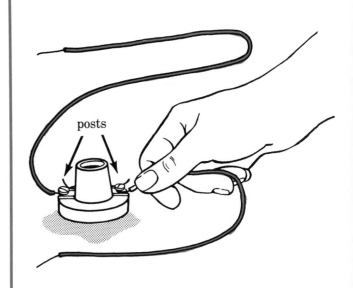

posts

3 Put the bulb into the socket.

4 Put the battery on top of one of the wires. It should cover the metal end of the wire.

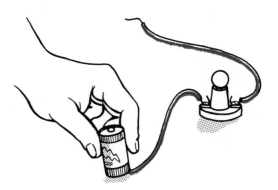

5 Pick up the other wire. Touch the top of the battery with it. Make sure the metal part of the wire touches the metal part of the battery.

Does the bulb light up? It should!

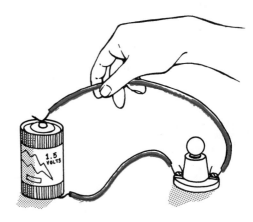

What Happens?

1. When does the bulb light up?
2. What energy makes the bulb light up?
3. Where does that energy come from?
4. The bulb changes the energy from the battery into another kind of energy. What kind is it?

Electricity from a Generator

Another way to get electricity is this: Use a **generator**. A generator is a machine that produces electricity. It changes motion energy into electrical energy.

You can see how a generator uses motion energy to produce electricity. You'll need the bulb, wires, and socket you used before. Put them together as you did for pages 24–25.

You'll also need a bicycle generator. (That's a small generator that fits on a bicycle. It lights up the bicycle headlight.) It should have two wires on it. It should also have a **rotor wheel**—a part that turns.

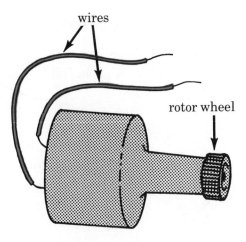

Bicycle generator

1 First join the wires on the generator to the wires on the socket. Twist the metal ends of the wires together.

2 Now twirl the rotor wheel as fast as you can. If you twirl fast enough, the bulb lights up.

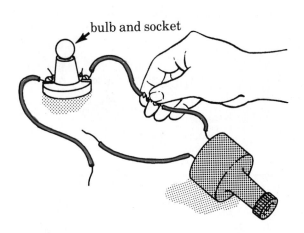

What Happens?

1. The generator changes motion energy into electricity. What gives motion energy to the generator? (Hint: What turns the rotor wheel?)
2. Suppose the generator is on a bicycle. What do you think turns the rotor wheel?

Inside a Generator

What does a generator have inside of it? Look at the diagram. What parts are inside?

Right! A coil of wire and magnets are inside the generator.

Electricity can be made in several ways. You know of one way—with chemicals in a battery. Another way is with magnets. Magnets have a force called **magnetism**. When you put two magnets close to each other, the magnetic force between them is very strong.

If a coil of wire moves through the magnetic force, this happens: Electricity is made in the wire. It flows through the wire.

Look at the diagram again. The coil of wire is wrapped around a metal part. That part is joined to something that sticks out of the generator. What is it?

Right! The metal part is joined to the rotor wheel.

What happens to the metal part when you turn the rotor wheel?

What then happens to the coil of wire?

Right! When you turn the rotor wheel, it turns the metal part. That moves the coil of wire. The wire moves through the magnetic force—and electricity is made!

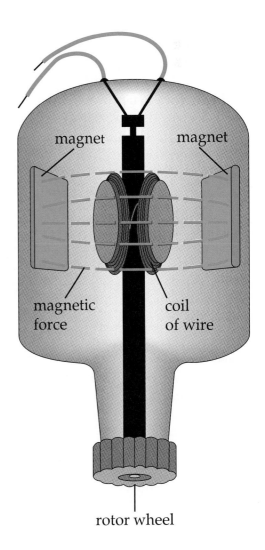

magnet   magnet

magnetic   coil
force   of wire

rotor wheel

Electricity from Running Water

Most of the electricity we use comes from **power plants**. Huge generators there make electricity. Those generators are like the bicycle generator you learned about. But they are much bigger.

A generator in a power plant works the same way that a bicycle generator does: Something turns a rotor wheel. The rotor wheel then turns a part that has a coil of wire. The wire moves between magnets. And electricity is made.

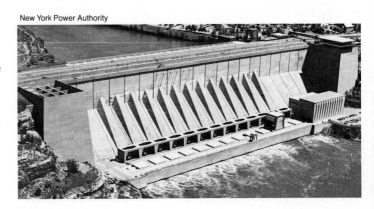

New York Power Authority

The picture on this page shows a hydroelectric power plant. *Hydro* means "water." What do you think moves the rotor wheels in a hydroelectric power plant?

The diagram on this page shows how a hydroelectric plant works. The plant is close to a dam. The dam makes a big lake. It controls the water that flows from the lake.

Some of the water goes through a big tube. The tube leads to the rotor wheel of a generator. The water has a lot of force. The water pushes the rotor wheel and turns it. The generator then makes electricity.

How does that electricity get to your house?

Right! Electricity travels from the power plant to your house through power lines.

On the diagram, find the rotor wheel, the generator, and the power lines.

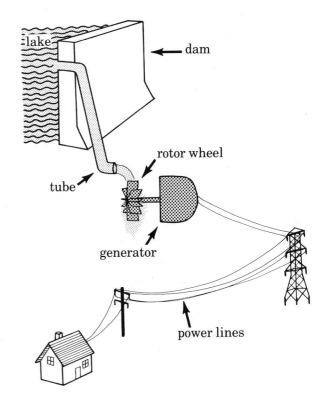

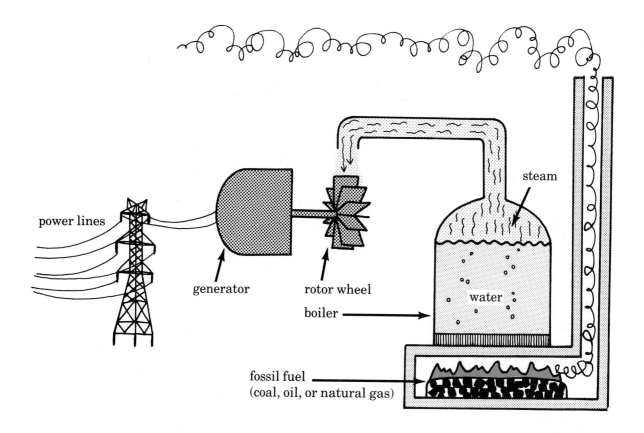

Electricity from Steam

Picture this:

You're boiling a pot of water. A cover is on the pot. The heat is turned up high. Suddenly, something pushes the cover up. What's pushing the cover?

Right! Steam is pushing the cover.

Steam has a lot of *pressure*. Pressure is a kind of force that can move things. Steam can have enough pressure to move a rotor wheel. So some power plants use steam to make electricity.

The diagram on this page shows how a power plant uses steam. First a lot of water is heated. Where is that water heated?

The water in the boiler is heated by a burning fuel. What kind of fuel is it?

When the water boils, it turns to steam. The steam goes through a pipe. The pipe leads to a rotor wheel. What happens next to make electricity?

How does the electricity get to a person's house?

Meet Your Meter

The electricity that comes to your house is made by an electric company. You must pay the company for all the electricity you use. How does the electric company know how much electricity you use? It sends someone to read your **meter**.

A meter is something that measures the units (amounts) of electricity people use. A unit of electricity is called a *kilowatt-hour*.

The picture shows what an electric meter looks like. The meter has four dials. You read the meter by looking at each dial.

Look at the first dial. The hand in the dial points to 3. So we read the first number as **3**.

Look at the second dial. What number does the hand point to?

Right! It points to 1. So we read the second number as **1**.

Look at the third dial. The hand is between two numbers. When that happens, we read the smaller number. What number is that?

Right! The smaller number is 2. So we read the third number as **2**.

Look at the fourth dial. What number should we read?

Now read all four dials again. How many kilowatt-hours does the meter show?

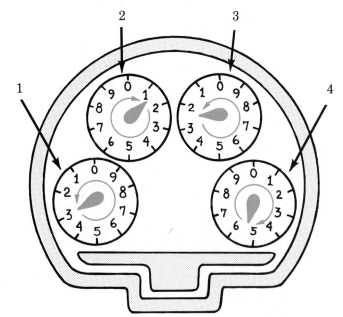

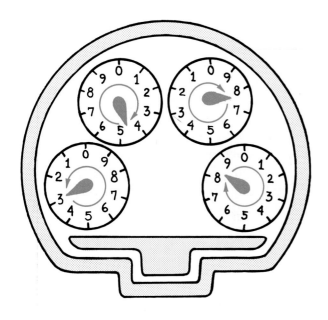

How Much Electricity?

The electric company sends you a bill every month. The bill shows you how much electricity you used during the month. And it shows how much money you owe.

Suppose the electric company wants to figure out how much electricity you used this month. This is what happens: It sends someone to read your meter at the end of the month. The number on the meter will be higher than it was last month. Why do you think that's so?

The meter keeps adding on the electricity you use. The dials on the meter move a little every time electricity is used in your house. So the meter shows a higher number every day than it showed the day before.

How does the company then figure out exactly how much electricity you used in a month?

Right! The company subtracts last month's number from this month's number.

Let's say the meter on this page shows the number for this month. And let's say the meter on the other page shows the number for last month. Read the meter for this month. On a separate piece of paper, write the number of kilowatt-hours it shows.

Now figure out how much electricity you used. Write the number for last month under the number you wrote for this month. Subtract last month's number from this month's number. Then check your answer. (The right answer is upside down.)

Answer

3478	this month	
−3125	last month	
353	amount used this month	

Review

Show what you learned in this unit. Finish the sentences. Match the words on the left with the correct words on the right.

1. A battery makes electricity
2. A generator makes electricity
3. A rotor wheel moves
4. A hydroelectric power plant
5. Some power plants use steam
6. An electric meter measures

a. a coil of wire inside a generator.
b. with motion energy.
c. with chemical energy.
d. how much electricity you use.
e. gets motion energy from water.
f. to turn the rotor wheels.

Check These Out

1. Find the electric meter at your house, apartment, or school. Read the meter two days in a row. Subtract the first number from the second number. How many units of electricity were used?
2. Get a flashlight. Open it up. Find the batteries inside. Find the bulb. Draw a diagram to show how the batteries and bulb work together.
3. Invite someone from the electric company to talk to your class. Ask that person about how electricity is made in your area.
4. Bring different kinds of magnets to class. Find out what they do.
5. Ask a car mechanic to show you a generator in a car. Ask that person to explain what it makes electricity for. Find out what turns the rotor wheel.
6. Bring an electric bill to class. Learn how to read it. How much does one unit of electricity cost? What different things does the bill show?
7. Here are more things you may want to find out:
   - Sound is a kind of energy. How does a stereo turn electrical energy into sound energy?
   - How does an electric motor work?
   - What is an electromagnet?
   - What is an electric circuit? How does it work?
   - What are electrons? What do they have to do with electricity?
   - How do magnets work?

Unit 5

Solving Energy Problems

We burn fossil fuels in our homes, cars, and factories. We also burn fossil fuels to make electricity. But burning fossil fuels causes problems.

- What are the problems with burning fossil fuels?
- What other ways can we get energy?
- How can we use less energy?

You'll learn the answers in this unit.

Before You Start

You'll be using the science words below. Find out what they mean. Look them up in the Glossary. On a separate piece of paper, write what the words mean.

1. **exhaust**
2. **pollute**
3. **radioactive**

Too Much Smoke!

Think about a car that someone is starting up. Smoke pours out of the car's exhaust pipe. What is happening in the engine to make the smoke?

Gasoline is burning inside the engine. As it burns, it produces smoke. The smoke comes out of the exhaust pipe. That smoke has different gases in it. It is also full of tiny bits of material.

Thousands of cars are driven every day. They all send out smoke. Thousands of factories and homes burn fossil fuels. They send out smoke too. Cars, factories, homes—all those make a lot of smoke. Where does the smoke go?

Right! It goes into the air. We breathe that air.

Fossil fuels give off lots of smoke and gases while they burn. So burning too much fossil fuel can pollute our air. For example, *smog* is made from smoke and other exhaust gases that come from cars, factories, and homes. Smog is a kind of air pollution. From far away, it looks like a heavy layer of dirty yellow or brown air.

What's bad about polluted air?

Yes! Polluted air can make people sick. If people breathe too much polluted air, it can hurt their eyes, throat, and lungs. It can even kill people who have breathing problems. Polluted air also harms trees, farm crops, and animals.

We can cut down the amount of pollution that goes into our air. How do you think we can do that?

Digging Deeper

Air pollution is just one problem with fossil fuels. Here's another problem: Fossil fuels near the surface of Earth are being used up. They are becoming harder to get.

Oil and coal companies must dig deeper into the ground to find fossil fuels. They must spend more money on workers and machines. So digging up fossil fuels has become expensive for those companies. How do you think that affects you?

Right! You have to pay more to use the products of fossil fuels.

Everything that comes from fossil fuels is becoming expensive. As fossil fuels become harder and harder to get, their prices go higher and higher. What are some things that will cost you more?

You will have to pay more for gasoline to run your car. Your gas and electric bills will be higher. White gas, kerosene, and all other products from fossil fuels will become expensive.

Some day it may be too hard and too expensive to get fossil fuels. People won't want to dig for them. It will be the same as running out of fossil fuels!

So people are starting to use other ways to get energy. What are some ways to get energy *without* using fossil fuels?

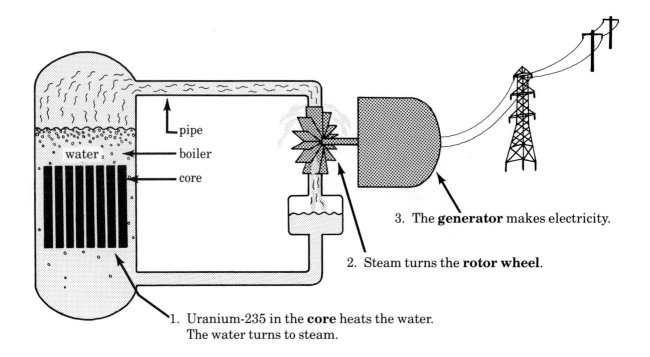

water

pipe

boiler

core

3. The **generator** makes electricity.

2. Steam turns the **rotor wheel**.

1. Uranium-235 in the **core** heats the water.
   The water turns to steam.

Nuclear Energy

Many power plants burn fossil fuels to make electricity. But some power plants are now using **nuclear fuel**.

Nuclear fuel is made from a special rock called **uranium-235**. Uranium-235 can release a lot of heat energy. It doesn't have to burn to give heat. So it doesn't make smoke.

Look at the diagram on this page. It shows how a nuclear power plant makes electricity.

1. The uranium-235 heats the water in the boiler. The uranium is in a special part. What is that part?
2. Steam goes through a pipe and turns a part. What is that part?
3. What then makes electricity?

Uranium-235 does not pollute our air with smoke. But it can cause other problems. It is *radioactive*. It turns into waste materials that are also radioactive. Radioactive materials can cause cancer in living things.

People may be harmed if radioactive materials are accidentally released. So nuclear power plants must be built very carefully. Waste materials must be put in special containers. Those containers must be moved very carefully and stored in safe places.

What problem might nuclear energy solve?

What problem might nuclear energy cause?

Wind Energy

Fossil fuels and nuclear energy cause problems. So people look for sources of energy that are cheap and *clean*. (Clean energy doesn't pollute.)

One source of energy that is cheap and clean is the wind. Wind power has been used for thousands of years. What is one way people have used the wind to move things on water?

Right! For thousands of years, people have used wind to sail ships. They have used wind energy to get motion energy.

People are now using wind energy to get electrical energy. Look at the picture on this page. It shows a special kind of windmill that makes electricity. The windmill has a generator and a rotor wheel. It is called a **wind generator**. How do you think it works?

When the wind blows, it turns the rotor wheel. As the rotor wheel turns, the generator makes electricity. The rotor wheel is shaped like an airplane propeller so it will turn easily in the wind.

Wind generators have one big problem. What do you think that problem is?

Right! The wind doesn't always blow. The generator can't make electricity when there is no wind. It makes electricity only on windy days. So some of the electricity that the wind generator makes is stored. It is stored in big batteries that are used on days when there is no wind.

rotor wheel

generator

NASA

solar cells

Electricity from the Sun

Heat energy can be changed into electricity. So can motion energy. What's a third kind of energy that can be changed into electricity?

Light energy can also be changed into electricity. We get that light energy from the sun.

People use **solar cells** to get electricity from sunlight. Solar cells are made from a special material called **silicon**. The silicon is cut into flat, round pieces. The pieces are the size of a small dish. Each side of the silicon is coated with a different chemical. A wire is connected to each side of the solar cell. When the sun shines on the silicon, electricity flows through the wire.

One solar cell doesn't produce enough electricity. So several solar cells are connected. Together they make enough electricity to run many appliances.

Solar energy, like wind energy, has a big problem. What do you think it is?

Right! The sun doesn't always shine. Solar cells can't make electricity at night or on cloudy days. So some of the electricity that's made on sunny days must be stored. It is stored in batteries to be used when there is no sunlight.

How do solar energy and wind energy help solve the problem of fossil fuels running out?

Right! The sun and the wind will always be around. They are **renewable** sources of energy. They will never run out.

How do solar and wind energy help solve the problem of pollution?

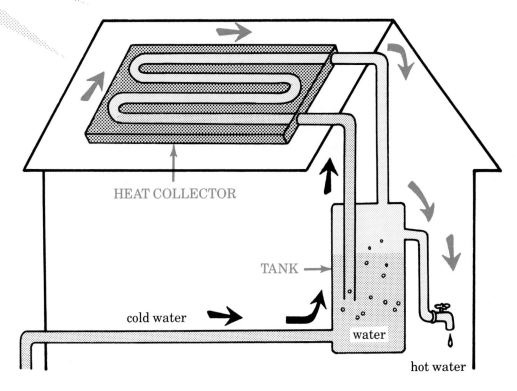

HEAT COLLECTOR

TANK

cold water

water

hot water

Heat from the Sun

There's another way to use the sun's energy. How is the water heated in your house, school, or place of work? Most people have water heaters that run on gas or electricity. But some people have *solar water heaters*. They use the sun to heat their water—and to save money.

Look at the diagram. It shows how a solar water heater works. The heater is on the roof of a building. One part of the heater is a large, flat piece of metal. What is the name of that flat metal part?

Yes! The flat metal part is a **heat collector**. It gets very warm when the sun shines on it.

The heat collector can get hot if it is painted a certain color. What do you think that color should be: black or white?

Things that are black get hot much faster than things that are white. (That's why black cars feel warmer than white cars when you touch them on a sunny day.) So heat collectors are painted black.

Look at the diagram again. Notice the long pipe that's joined to the heat collector. That pipe is a water pipe. When water flows through the pipe, the water picks up heat from the heat collector. The hot water then flows to a part where it is stored. What part stores the hot water?

The tank stores the water that's heated by the sun. That water can get as hot as water heated by gas or electricity!

Experiment 2

Which material makes a good insulator?

One way to solve energy problems is to use less energy. For example: If we keep hot things from cooling off, we won't have to heat them up again. And we save energy.

If you put hot things in certain materials, you can keep them from losing their heat. Those materials are called **insulators**. Does metal or Styrofoam make a good insulator?

Materials

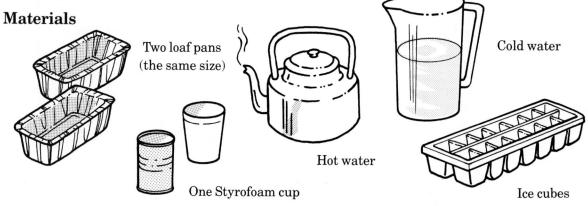

Two loaf pans (the same size)

Hot water

Cold water

Ice cubes

One metal cup (or metal can)   One Styrofoam cup (the same size as the metal one)

Procedure

1. Place the Styrofoam cup in one loaf pan. Put the metal cup in the other loaf pan.

2. Put half the ice cubes in one pan. Put the other half in the other pan. (Do not put ice in the cups.)

3. Pour hot water into one cup. Pour the same amount of hot water into the other.

4. Fill both loaf pans with cold water. Wait about five minutes. Then touch the water in one cup. Next touch the water in the other cup.

Observations

1. Is the water hotter in the Styrofoam cup or the metal cup?
2. Look at the ice cubes in the two pans. Did the ice melt faster around the Styrofoam or the metal cup?

Conclusions

1. Which material let out more heat: the metal or the Styrofoam?
2. Which material was better at keeping the water hot?
3. Which material makes a good insulator?

Review

Show what you learned in this unit. Match the words in the list with their meanings. The page number after each meaning tells where you can find the word.

nuclear radioactive clean
insulator pollute   wind
exhaust   solar

1. What we call energy that doesn't dirty the air (page 37)
2. A material that insulates (page 40)
3. Energy from air that moves (page 37)
4. Smoke that comes out of an engine (page 34)
5. Energy from uranium-235 (page 36)
6. The kind of waste materials that uranium turns into (page 36)
7. Energy from the sun (page 38)
8. To make air dirty (page 34)

Check These Out

1. One way to use wind energy is with a sail. Make model sailboats that have sails of different sizes. What kind of sail moves the boat fastest?
2. Invite someone from an ecology group to come to your class. Ask that person how your community can solve energy problems.
3. Some calculators use solar cells instead of batteries. Borrow a solar calculator. What happens when you cover the solar cell?
4. Here are more things you may want to find out:
   - Air pollution makes people sick in many ways. What are some of those ways?
   - How does uranium-235 become hot? Find out about nuclear chain reactions.
   - What are these other ways to get energy: biomass, coal gasification, shale oil?
   - What makes smog? What is photochemical smog? What's bad about smog? How can it be controlled?
   - What is acid rain? What causes it? What's bad about acid rain? How can it be controlled?
   - What is the greenhouse effect? What does burning fossil fuels have to do with it?

Unit 6

What Can You Do?

Most of the energy we use comes from fossil fuels. But fossil fuels cause problems! They pollute the air we breathe. And they are expensive. Our gas, electric, and fuel bills keep getting higher and higher.

What can we do about the problems of fossil fuels? We can **conserve** energy. That means we can use fossil fuels wisely. We can use them only when we need them and not waste them.

What things do you use that run on fossil fuels? Answer the questions below.

1. What are some things that use gasoline or diesel fuel?
2. What are some things that use electricity?
3. What are some things that use natural gas?
4. What are some things that use kerosene or white gas?

Think of some ways you can conserve energy with the things you use.

Save on Gasoline

Here are some ways you can conserve gasoline and save money:

1. Walk or ride a bicycle instead of driving a car.

2. Take a bus or subway instead of driving a car.

3. Form a carpool. Ride with others to work or school. Use just one car.

4. Never drive faster than 55 miles per hour!

What other ways can you conserve gasoline?

Save on Electricity

Here are some ways you can conserve electricity and save money:

1. Turn off lights when you leave a room.

2. Turn off the TV when you are not watching it.

3. Turn off the stereo when you are not listening to it.

4. Don't use an appliance unless you really need to.

What else can you do to conserve electricity?

Save on Heat

Here are some ways you can conserve heat and save on your heating bill:

1. Keep the doors closed. This will keep the heat inside.

2. Keep the windows closed. Cover them with curtains or shades.

3. If you are cold, put on a sweater. Don't turn up the heat.

4. Turn off the heater when you won't be home.

What else can you do to save on your heating bill?

Save Energy at Home

Here are more ways you can save energy at home:

1. Keep the refrigerator door closed as much as you can.

2. When you are cooking, keep pots covered. Keep the oven door closed when the oven is on.

3. Run your dishwasher and clothes washer only when they are full.

4. Hang your clothes to dry instead of using a dryer.

How else can you conserve energy at home?

Show What You Learned

What's the Answer?

Find the answers for the questions. There may be more than one correct answer to a question.

1. What are kinds of energy?
   a. Heat, light, motion
   b. Car, battery, grill
   c. Lamp, stove, radio

2. What happens to potential energy?
   a. It pollutes the air.
   b. It is made.
   c. It is changed to heat, light, and motion.

3. What changes the potential energy in fossil fuels to heat, light, and motion?
   a. Mining
   b. Burning
   c. Conserving

4. Which are used to make electricity?
   a. Steam
   b. Running water
   c. Wind

5. Where is electricity made?
   a. In batteries and generators
   b. In solar cells and power plants
   c. In refineries and coal mines

6. What renewable sources of energy will never run out?
   a. Wind, sun
   b. Crude oil, natural gas
   c. Coal

What's the Word?

Give the correct word or words for each meaning.

1. Something we burn to get heat, light, or motion
   F _____

2. Energy that is stored
   P _____

3. A machine that makes electricity
   G _____

4. To make the air dirty
   P _____

5. Something that changes the sun's energy into electricity
   S _____ C _____

6. A material that keeps heat from leaving something
   I _____

7. To use energy wisely
   C _____

Congratulations!
You've learned a lot about energy. You've learned

- What kinds of energy we use
- How we get energy
- How we can conserve energy
- And many other important facts about energy

ELECTRICITY

What kind of force is electricity? How do we use electricity? How does electricity travel? How can we use electricity safely? In this section, you'll learn many facts about electricity. And you'll learn about the important part electricity plays in our lives.

Contents

Introduction

Picture this:

It's early in the morning. You're asleep. Your alarm clock wakes you.

You wash your hair and then dry it with a hair dryer. You make some toast. The radio is on while you eat your breakfast.

You leave the house and head for the bus stop. You listen to your transistor radio until the bus comes along. As you ride along, the traffic lights give their red, yellow, and green signals.

During the day, you buy some things. The store clerks use machines to ring up the sales. You make some phone calls.

At home that evening, you watch TV. A friend rings the doorbell. You turn on a lamp. You put your dinner in the oven. Later you set your alarm clock. You turn out the light and go to bed.

All day long you use electricity. You depend on it to wake you and to cook your food. You use it to play music, to go from place to place, and for many other things.

Electricity is an important part of your life. In this section, you'll learn how electricity works and how to use it safely.

Unit 1

A Special Path

Did an electric alarm clock wake you today? Did you turn on a lamp? Use an electric hair dryer?

If you did, you used electricity. Electricity is a kind of energy that can make things work. But for electricity to make things work, it must have a special path that it can follow.

- What is that path called?
- What is that path like?

You'll learn the answers in this unit.

Before You Start

You'll be using the science words below. Find out what they mean. Look them up in the Glossary that's at the back of this book. On a separate piece of paper, write what the words mean.

1. **battery**
2. **flow**

There and Back Again

Click! You turn on a lamp and the **bulb** lights up. Why does that happen?

The bulb lights up because electricity flows through it. For electricity to flow, it needs a special path to follow. That path is called a **circuit**.

What's a circuit like? Look at the diagram. With your finger, trace a line that connects all the dots to each other.

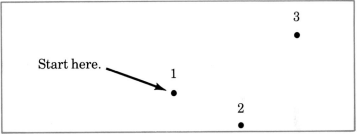

Imagine that the line you traced is a wire that electricity flows through. Trace the path of the wire again. Start at **1**. (**1** is like a place where electricity is made.) Move your finger to **2** and then to **3**. (**2** and **3** are like a lamp.)

Where do you go from **3**?

Right! You go back to **1**. That's what an electric circuit is like. It starts at a place where electricity is made. It goes away from that place. Then it goes back to that place.

The electricity you use every day is probably made at a place many miles from your home. It's made by a huge machine.

But electricity can also be made by other things. In fact, you may have one of those things in your home. You use it to make a flashlight work. What do you think it is?

Make a Circuit

You can make a circuit with things from a hardware store. These are the materials you will need:

- Two pieces of bell wire (each 12 inches long)
- One sharp knife
- One very small socket (with 2 posts)
- One 6-volt battery (it has 2 posts)
- One small bulb (like a flashlight bulb) that fits into the socket
- One wire stripper or sharp knife

Caution: Be careful when you use a knife.

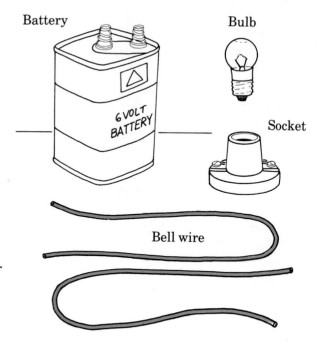

Battery   Bulb   Socket

6 VOLT BATTERY

Bell wire

---

1 Strip the ends of each wire. First, carefully cut the covering one inch from the end. (Don't cut through the wire.) Cut all around the wire.

Pull the covering off the wire. The end will then be stripped.

2 Get the socket. Connect a stripped end of one wire to a post.

Connect a stripped end of the other wire to the other post.

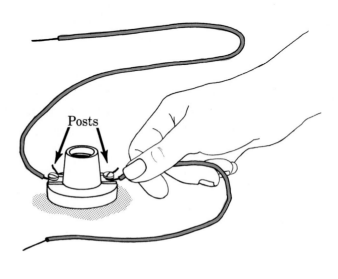

Posts

3 Get the battery. Pick up one of the wires that's connected to the socket. Connect the stripped end of that wire to a post on the battery.

Connect the stripped end of the other wire to the other post on the battery.

4 Screw the bulb into the socket. You've just made a circuit.

Did the bulb light up? It should!

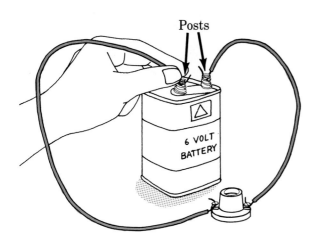

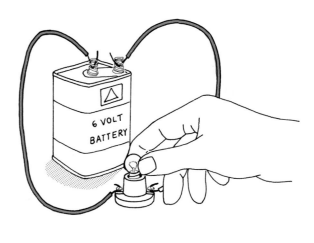

What's Happening?

1. What makes electricity in your circuit?
   Yes! The battery makes electricity.
2. What makes a path for the electricity to follow?
   That's right! The wires make a path.
3. What happens to the bulb when electricity flows through your circuit?

Experiment 1

What happens when you open a circuit?

You made a circuit with a bulb in it. The bulb lit up. It lit up because electricity flowed through it.

Electricity flowed because the circuit was closed. The circuit had no opening in it.

What happens if you open the circuit? Do the experiment and find out.

Materials (What you need)

- Your circuit (set up so that the bulb is lit up)

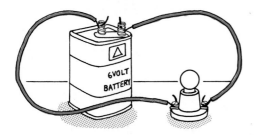

Procedure (What you do)

1. Open the circuit by doing this: Take the wire off one of the battery posts. What happens to the bulb?

2. Close the circuit by doing this: Connect the wire to the battery post again. What happens to the bulb now?

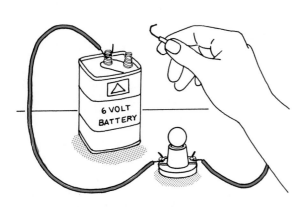

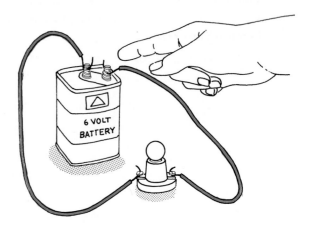

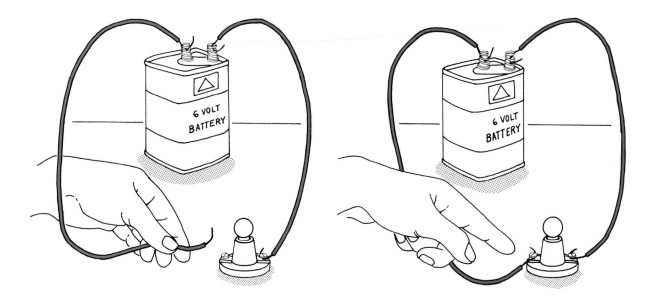

3. Open the circuit again. This time, take the wire off one of the socket posts. What happens to the bulb?

4. Close the circuit. Connect the wire to the socket post again. Now what happens to the bulb?

Observations (What you see)

1. What happens to the bulb when you open the circuit?
2. What happens to the bulb when you close the circuit?

Conclusions (What you learn)

What happens when you open a circuit? Does electricity flow or stop flowing?

Review

Show what you learned in this unit. Match the words in the list below with their meanings.

battery   electricity   bulb
circuit   closed   flows

1. A kind of energy
2. A circuit with no opening in it
3. Something that makes electricity
4. How electricity moves
5. A path for electricity
6. Something that lights up

Check These Out

1. Make a Science Notebook for Electricity. Keep your list of glossary words and their meanings in the notebook. Also keep a record of the experiments you do. Put anything else you learn about electricity in your Science Notebook.
2. On page 53, you read about a "huge machine" that makes electricity. That machine is called a generator. Find out how it works.
3. Find out how a battery makes electricity.
4. Make a list of the things in your home that run on batteries.
5. As you work through this section, you may want to find out more about electricity. You can find out more by looking in a dictionary or an encyclopedia, or by getting books about electricity from a library. You can also talk with an expert, such as an electrician, an electrical engineer, a clerk in a hardware store, or a science teacher.

     Here are some things you may want to find out:
   • What are atoms? What are electrons?
   • What is static electricity? What is a positive charge? What is a negative charge?

Unit 2

On and Off

Think about a lamp you use at home. Do you have to connect a wire each time you turn the lamp on? No. You just use the lamp switch. And you use the same thing to turn the lamp off.

- How does a switch turn a lamp on?
- How does a switch turn a lamp off?

You'll learn the answers in this unit.

Before You Start

You'll be using the science words below. Find out what they mean. Look them up in the Glossary. On a separate piece of paper, write what the words mean.

1. **control**
2. **switch**

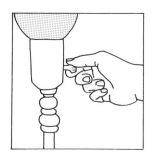

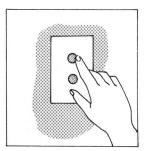

Different kinds of switches

You're in Control

Can you control electricity? You do it every time you turn a lamp on or off. You do it by using the lamp switch.

A switch lets you control electricity without handling any wires. It lets you start and stop electricity that flows through things such as lamps, radios, and machines.

There are many kinds of switches. Some are buttons that you push. Others are chains that you pull. Still others are knobs that you turn. But all switches work the same way.

How does a switch work? Look at the diagram at the right. It shows an open circuit. Because the circuit is open, electricity can't flow. So the lamp (bulb) is not on.

Suppose you take wire **A** and touch its end to wire **B**. When you do that, you connect **A** and **B**, and you close the circuit. Electricity flows and the lamp goes on.

Now suppose you take wire **A** *away* from wire **B**. You open the circuit. Electricity can't flow. Is the lamp on or off when the circuit is open?

Right! It's off.

That's how a switch works. It lets you close a circuit so that electricity can flow. And it lets you open a circuit when you want to stop the electricity from flowing.

Diagram of an open circuit

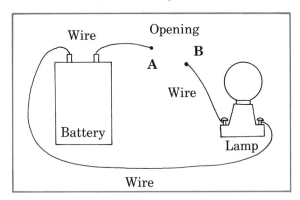

Open or Closed?

The two diagrams on this page show how a switch opens and closes a circuit. Look at the diagrams and answer the questions.

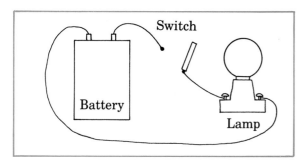

Diagram A

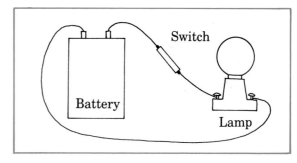

Diagram B

1. Which diagram shows the switch closing a circuit?
2. Which diagram shows the switch opening a circuit?
3. Look at diagram **A**. Is the lamp on or off?
4. Look at diagram **B**. Is the lamp on or off?
5. Now finish these sentences with *closes* or *opens*.
   a. When the switch _____ the circuit, the electricity stops flowing and the lamp is off.
   b. When the switch _____ the circuit, the electricity starts flowing and the lamp is on.

Make a Switch

Here's an easy switch to make. You'll need these materials:

- One piece of cardboard (about 4 inches long and 2 inches wide)
- Two short pieces of bell wire (each 6 inches long)
- Two 1-inch brass fasteners
- One 2-inch paper clip
- The circuit you made for Unit 1 of this section (with its battery)

Make sure the ends of the short wires are stripped. If they aren't, take off 1 inch of covering from each end.

Now take one of the wires out of your circuit. Take it out completely.

Make a switch for the circuit. Follow these directions.

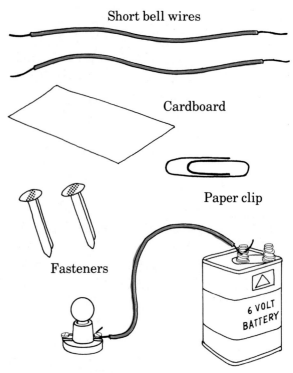

Short bell wires

Cardboard

Paper clip

Fasteners

6 VOLT BATTERY

Circuit with only one wire on it

---

1 Make two holes in the cardboard. (Use a paper punch or poke out the holes with a pencil.) The holes should be about 1 inch apart.

2 Push a fastener through a hole. Push the other fastener through the other hole.

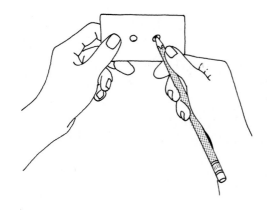

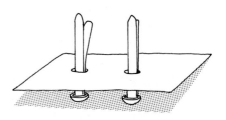

3 Wrap one end of a short wire tightly around a fastener. (Make sure the stripped part touches the fastener.) Spread the fastener open.

Do the same thing with the other wire and fastener.

4 Turn the cardboard over. Hook the paper clip over the head of a fastener. Write *OFF* near that fastener. Write *ON* near the other one.

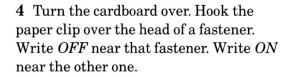

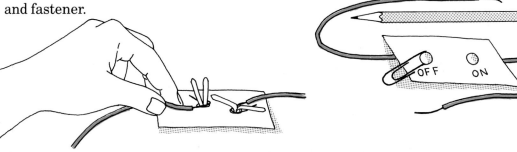

5 Now add the switch to the circuit. First, connect one switch wire to the empty post on the socket.

6 Then connect the other switch wire to the empty post on the battery.

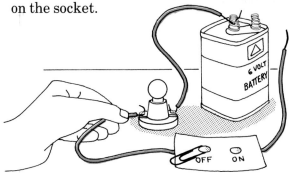

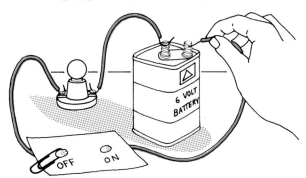

7 Turn the switch on. Move the paper clip so that it touches the heads of both fasteners.

8 Does the bulb light up? (If it doesn't, check the wires, the fasteners, and the paper clip. Make sure they're all touching.)

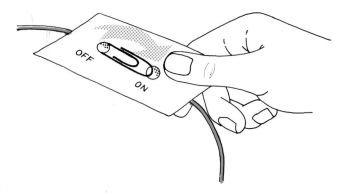

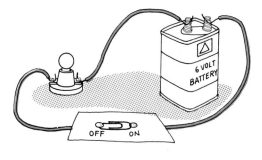

Review

Show what you learned in this unit. Finish the sentences. Match the words on the left with the correct words on the right.

1. You can use a switch
2. A switch opens or
3. Electricity flows through
4. Electricity can't flow through

a. closes a circuit.
b. an open circuit.
c. to control electricity.
d. a closed circuit.

Check These Out

1. Look around your home to see how many kinds of switches you can find. Draw a picture of each kind.
2. Get a wall switch. It can be a used one or a new one from a hardware store. Show the switch to the class. Explain how its parts work.
3. Here are more things you may want to find out:
   - What is current electricity?
   - What is voltage?
   - What is a transformer?

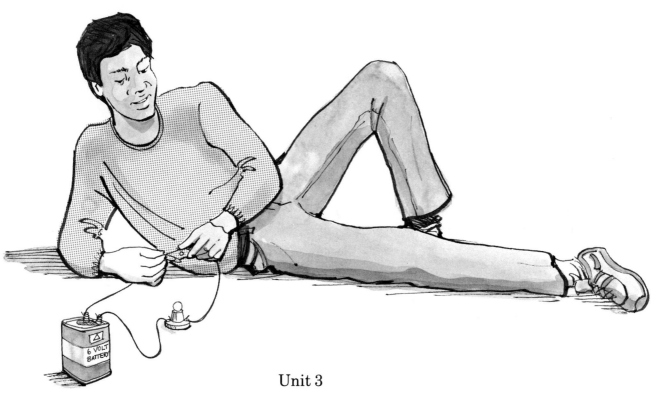

Unit 3

Go with the Flow

You've seen that electricity flows through the wires in your circuit. And you've seen that it flows through the paper clip in your switch.

But electricity *can't* flow through some things.

- What things can electricity flow through?
- What things can't electricity flow through?

You'll learn the answers in this unit.

Before You Start

You'll be using the science words below. Find out what they mean. Look them up in the Glossary. On a separate piece of paper, write what the words mean.

1. **conduct**
2. **shock**

Cover Story

Look at a piece of bell wire. It has two parts—a wire center and a covering. Which part conducts electricity?

Right! The wire center conducts electricity. In other words, electricity flows through it.

The center is made of metal. Metal conducts electricity. Anything that conducts electricity is a **conductor**. So metal is a conductor.

Anything that electricity *can't* flow through is a **nonconductor**. For example, look at the covering of the bell wire. It's made of something that doesn't conduct electricity. What is the covering made of?

What do you think could happen if the wire had no covering?

If the wire had no covering, electricity could flow out of the wire. It could flow into any conductor that touched the wire. For example, if *you* touched the wire, electricity could flow into you. You could get a shock.

Now look at the picture on this page. Find the conductor (the part of the wire that conducts electricity). Find the nonconductor (the part of the wire that doesn't conduct electricity).

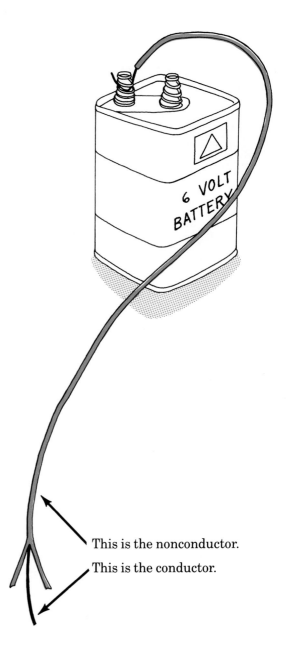

This is the nonconductor.

This is the conductor.

Which Is Which?

The pictures on this page show things you can use every day. Guess which things can conduct electricity. Guess which ones can't.

On a separate piece of paper, write down each thing's name. Write *conductor* next to each one you think can conduct electricity. Write *nonconductor* next to each one you think can't conduct electricity.

Metal penny

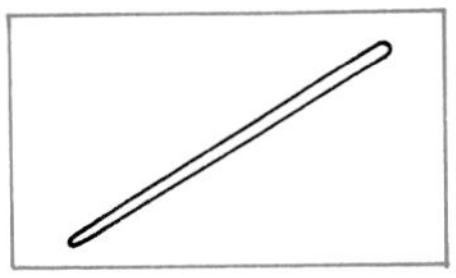

Wood toothpick

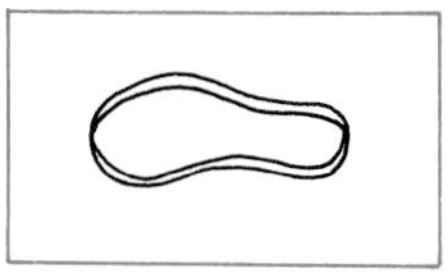

Rubber band

Metal nail

Glass

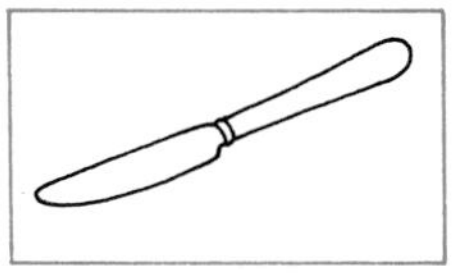

Metal knife

Metal key

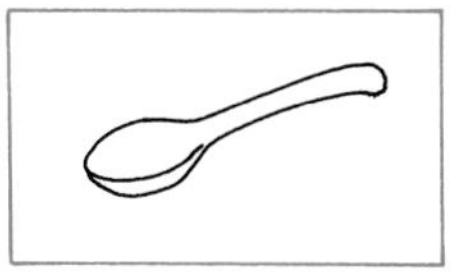

Plastic spoon

Find out if your guesses are correct. Turn the page and do the experiment.

Experiment 2
Which things are conductors?

Materials

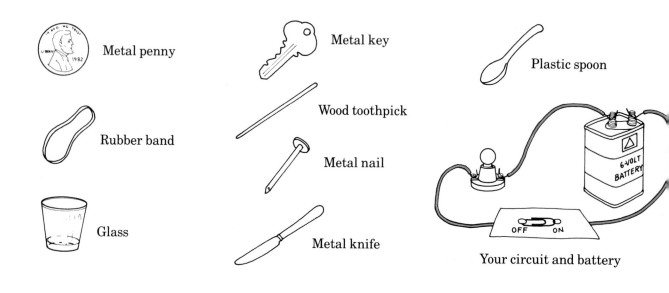

Metal penny

Metal key

Plastic spoon

Rubber band

Wood toothpick

Metal nail

Glass

Metal knife

Your circuit and battery

Procedure

1. Put everything on a table in front of you. If your circuit has a switch, unwrap the wires on the switch fasteners. Then take the switch off the circuit.

2. Get the penny. Pick up the two loose wires on the circuit. Touch the penny with the stripped ends of the two wires.

   Does the bulb light up? If it does, the penny is a conductor.

   Test the other things in the same way.

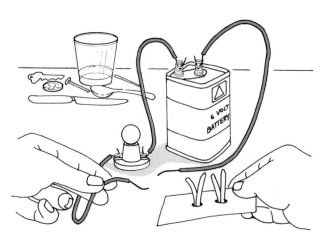

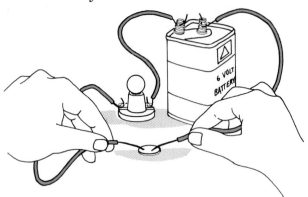

Observations

On a separate piece of paper, make a list of each thing that lights up the bulb. Next to each thing on the list, write what it is made of. For example, the penny lights up the bulb. Next to *penny*, you would write what it's made of: *metal*.

Then make a list of each thing that doesn't light up the bulb. Next to each thing, write what it's made of.

Conclusions

Find the right word or words in the list to finish each sentence.

1. Electricity can flow through _____.

   rubber   metal   glass   plastic   wood

2. Electricity can't flow through _____.

   rubber   metal   glass   plastic   wood

3. The things made of _____ are conductors.

   rubber   metal   glass   plastic   wood

How did you do?

Turn back to page 67. These things are conductors: penny, key, nail, knife.

These things are nonconductors: rubber band, glass, toothpick, spoon.

Review

Part I

Show what you learned in this unit. Match the words in the list below with their meanings.

conductor   nonconductor   conduct

1. To carry something such as electricity
2. Something that can't carry electricity
3. Something that can carry electricity

Part II

Match the clues on the left with the correct words on the right.

4. A nonconductor that windows are made of
5. A nonconductor that's used to cover wire
6. A nonconductor that tires are made of
7. A nonconductor that some chairs are made of
8. A conductor that nails are made of

a. rubber
b. wood
c. glass
d. plastic
e. metal

Check These Out

1. Look around your home. Find three things that are conductors and three things that are nonconductors. Bring them to class. Ask the class to tell which things are conductors and which are nonconductors.
2. Go to a hardware store and look at different kinds of electrical wires. Ask a salesperson these questions about the wires:
   - What different kinds of metals are used in the wires?
   - What different kinds of materials are used as coverings for the wires?
3. Here are more things you may want to find out:
   - What is an insulator?
   - What are free electrons?
   - Why do metals make good conductors?

Unit 4

Electric Light

Look at a light bulb when it's on. What do you see? Of course, you see light.

Now put your hand near the bulb. What do you feel? Yes, you feel heat.

The heat and the light are both made by electricity flowing through the bulb.

- Why does a bulb give off light?
- How does electricity flow through a bulb?

You'll learn the answers in this unit.

Before You Start

You'll be using the science words below. Find out what they mean. Look them up in the Glossary. On a separate piece of paper, write what the words mean.

1. **cord**
2. **glows**
3. **incandescent**

It's So Hot, It Glows

Look at the light bulb in your circuit. What's inside the light bulb?

Right! A very thin wire.

Now set up the circuit so that the light bulb is on. What part of the bulb glows?

Right again! The thin wire inside the bulb is the part that glows.

That wire is called a **filament**. The filament is made of a special metal that can be heated quickly and easily. When the metal is heated, it glows. In other words, it gives off heat and light.

Light bulbs that have a filament are called incandescent light bulbs. Most of the bulbs you use every day are incandescent bulbs. Incandescent bulbs are used in lamps and flashlights. They are used in cars, on Christmas trees, with cameras, and so on.

Incandescent light bulbs come in many shapes and sizes. But they all work the same way. They all have a filament that glows when electricity flows through it.

Look at the diagram at the right. It shows how electricity flows through a light bulb. The electricity goes into the metal bottom of the bulb. It flows through the filament. Where does it come out again?

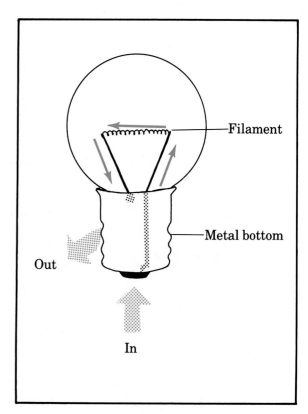

How electricity flows through a bulb

Back to the Circuit

What heats the filament in a light bulb?
Electricity does.

How does electricity get to and from the light bulb?
You already know these things:

- Electricity follows a path.
- That path is made of something that is a conductor.
- Metal is a conductor.

So you know that electricity gets to and from the light bulb by following a metal path.

Think about the circuit you made. You connected two metal wires to the metal posts on a battery. And you also connected those two wires to the metal posts on a socket. Then you added a bulb that has a metal bottom and a metal filament. You made a metal path for electricity to follow.

The picture below shows a circuit like the one you made. With your finger, trace the metal path that goes to and from the bulb.

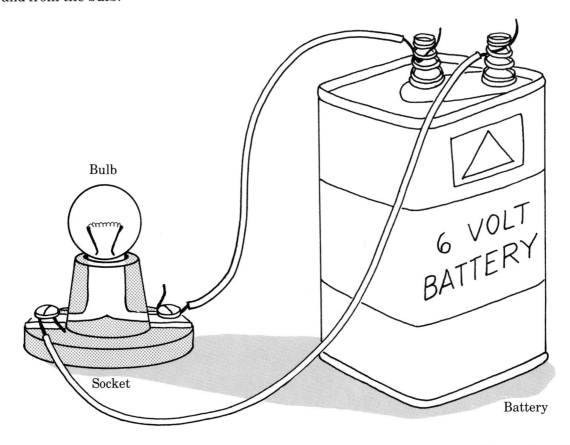

Bulb

Socket

6 VOLT BATTERY

Battery

Two-Way Traffic

Think of a lamp like the one in the picture. What conducts electricity to and from the lamp?

Right! The lamp cord.

How many wires do you think the cord has?

It looks as if it has only one wire. But it really has two. If you look closely at the cord, you'll see this: It's made up of two covered wires that are stuck together.

The two wires in the cord work just like the wires in your circuit. One cord wire conducts electricity one way—to the lamp. The other cord wire conducts electricity the other way—from the lamp.

With your circuit, you can see how a cord works. You'll need these materials:

- One cord with a plug on it (from a hardware store)
- Your circuit with its switch and battery

WARNING: This activity should be done with a battery only. *Do not use a wall outlet!* The electricity from an outlet is too powerful.

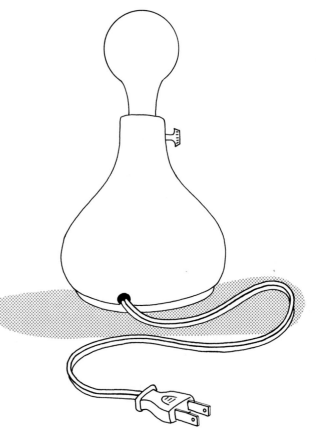

---

1 Pull the two wires of the lamp cord apart for a few inches. Strip the end of each wire. Then the cord should look like the picture.

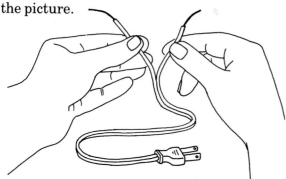

2 Get your circuit. Take both wires off the battery posts.

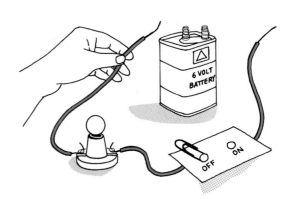

3 Pick up the stripped end of a cord wire. Twist it around the stripped end of the switch wire.

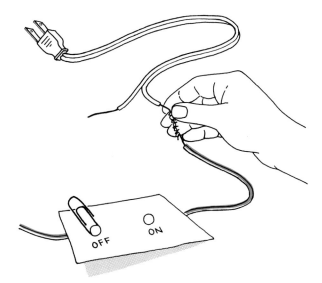

4 Pick up the stripped end of the other cord wire. Twist it around the stripped end of the socket wire. Make sure the switch is off.

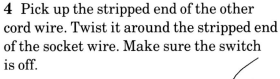
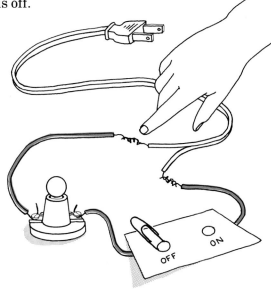

5 Pick up the plug. Press one metal part tightly against one of the battery posts. Press the other metal part tightly against the other battery post.

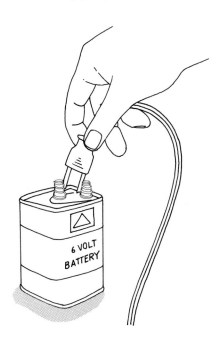

6 Turn on the switch. What happens?

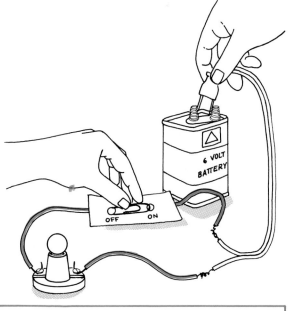

> **WARNING:** Don't fool around with electricity. It can hurt you. *Do not plug this circuit into a wall outlet.*

Review

Show what you learned in this unit. Finish the sentences. The words you'll need are listed below. (You'll use one of the words twice.)

metal   filament   incandescent
cord   glows

1. The thin wire in a light bulb is called a _____.
2. The filament is made of a special _____.
3. When the filament gets very hot, it _____.
4. An _____ bulb is a light bulb that has a filament.
5. There are two wires in an electric _____.
6. Electricity follows a _____ path.

Check These Out

1. You know that light can be made by a filament that glows. But some kinds of light are not made by a filament. One of these kinds of light is called *fluorescent* light. Find out how it is made.
2. Another kind of light that is not made by a filament is called *neon* light. Find out how it is made.
3. Find out how a toaster works. (Hint: It works something like a light bulb.)
4. Here are more things you may want to find out:
   - Who was Georg S. Ohm? What did he discover about current electricity?
   - What is an ohm?
   - What is resistance?
   - Some conductors have more resistance than others. Why?
   - How is resistance used in appliances that make heat, such as toasters?

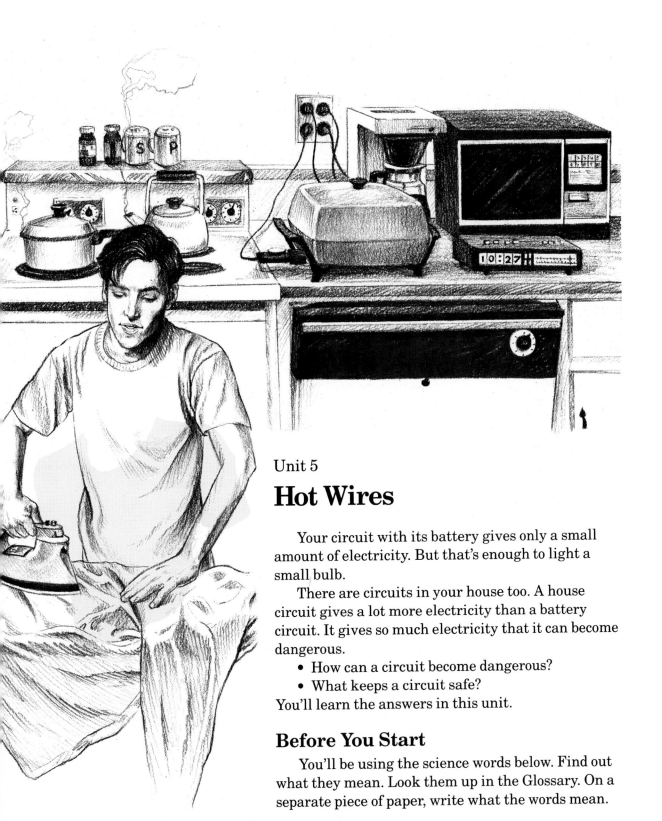

Unit 5

Hot Wires

Your circuit with its battery gives only a small amount of electricity. But that's enough to light a small bulb.

There are circuits in your house too. A house circuit gives a lot more electricity than a battery circuit. It gives so much electricity that it can become dangerous.

- How can a circuit become dangerous?
- What keeps a circuit safe?

You'll learn the answers in this unit.

Before You Start

You'll be using the science words below. Find out what they mean. Look them up in the Glossary. On a separate piece of paper, write what the words mean.

1. **appliance**
2. **outlet**

A Lot of Electricity

How do you connect a lamp or a toaster to a circuit in your house?

Right! You plug the lamp or toaster cord into a wall outlet. The wall outlet is part of the house circuit. That circuit can give a lot of electricity. How much can it give? That depends on what's plugged into the circuit.

Suppose you plug a toaster into an outlet and turn the toaster on. A lot of electricity flows through the circuit. A toaster needs a lot of electricity.

But suppose you plug a lamp into an outlet and turn the lamp on. Less electricity flows through the circuit. A lamp doesn't need as much electricity as a toaster.

Why do you think a toaster needs a lot of electricity? Get a toaster and find out.

Look inside the toaster. What do you see?

Right! You see a lot of wires.

Now plug in the toaster and turn it on. What happens to the wires?

Right! They get very hot—hot enough to toast bread. It takes a lot of energy to heat all those wires. That's why a toaster needs a lot of electricity.

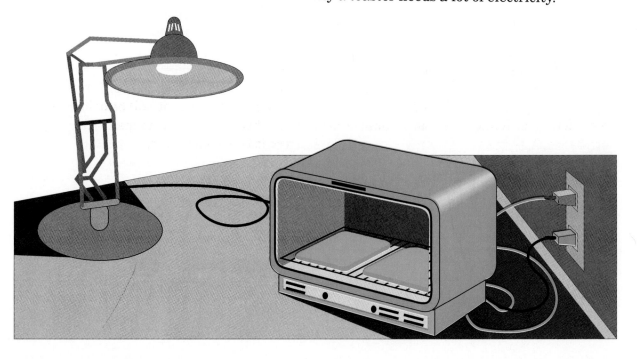

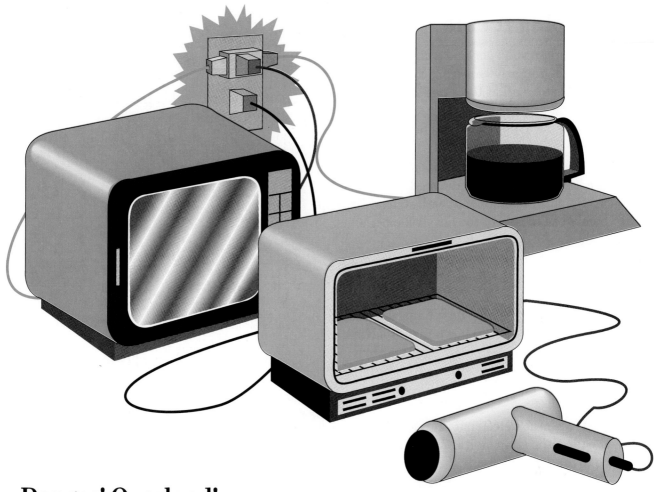

Danger! Overload!

You've seen that a toaster needs a lot of electricity to make heat. There are other appliances that make heat. And those appliances need large amounts of electricity too. What are some of those appliances?

Suppose you plug two or three of those appliances into an outlet and turn them all on. What happens in the circuit?

Right! Lots and lots of electricity flows. The more appliances you use at the same time, the more electricity flows.

If you use too many appliances on one circuit, the circuit wires will be **overloaded**. The wires will carry more electricity than they can safely handle.

Look at the picture above. It shows many appliances plugged into one outlet. If those appliances are all turned on at the same time, the circuit will be overloaded.

What could you do to make sure the circuit isn't overloaded?

That's right! Don't plug so many appliances into one outlet.

Keeping It Safe

Suppose you overload a circuit by mistake. You use your iron, coffeepot, and electric stove all at the same time. A lot of electricity flows. The circuit wires get warm . . . warmer . . . hot! They could start a fire!

Suddenly the stove, the toaster, and the iron go off. What happened?

Something stopped the electricity before the wires got too hot. It kept the circuit safe.

What stopped the electricity was a **fuse**. A fuse is like a switch. It can open a circuit.

The picture on the next page shows one kind of fuse. The fuse has a metal bottom, like a light bulb. That bottom fits into something like a socket. The fuse has a glass top. The top covers a thin strip of metal.

If too much electricity flows through the fuse, the strip melts. What does that do to the circuit?

Right! It opens the circuit.

What happens when the circuit is opened?

Yes! The electricity stops flowing. The circuit wires cool off. And the circuit is safe.

Each circuit in a house has a fuse. All the fuses are in a **fuse box**. (Your house may have **circuit breakers** instead of fuses. They work the same way.)

Where can you find the fuse box (or circuit breakers) in your house?

Flowing Through a Fuse

Get a fuse. (You can find one in a hardware store.) Look at the fuse. Do you see the metal bottom? The thin metal strip? Those are the two most important parts of the fuse.

Electricity flows through a fuse the same way it flows through a light bulb. It follows a metal path.

What part of a fuse does the electricity go into first?

Right! Electricity goes into the metal bottom.

What part of a fuse does electricity then flow through?

What part of a fuse does electricity come out of?

With your finger, trace the path of electricity in a fuse.

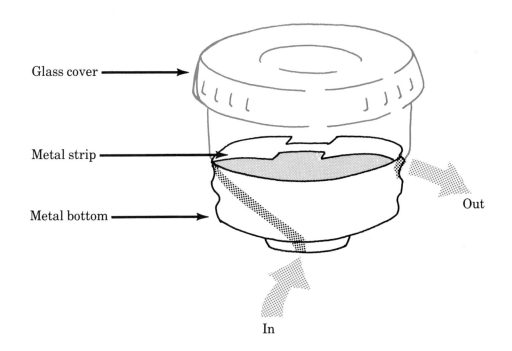

Glass cover

Metal strip

Metal bottom

Out

In

Review

Show what you learned in this unit. Finish the sentences. Match the words on the left with the correct words on the right.

1. Appliances that make heat need
2. A fuse can open a circuit
3. The strip in a fuse melts if
4. An overloaded circuit can

a. and keep it safe.
b. start a fire.
c. large amounts of electricity.
d. too much electricity flows through it.

Check These Out

1. Find out how many circuits there are in your house. Count the fuses in the fuse box. Or ask someone. Compare what you find out with what others in the class find out.
2. When the thin strip in a fuse melts, the fuse must be replaced. Find out how to replace a fuse safely.
3. How does a circuit breaker work? Find out and explain it to the class.
4. Here are more things you may want to find out:
   - What is an ampere? What is a watt?
   - What is direct current? What is alternating current?
   - How does an electric motor work?

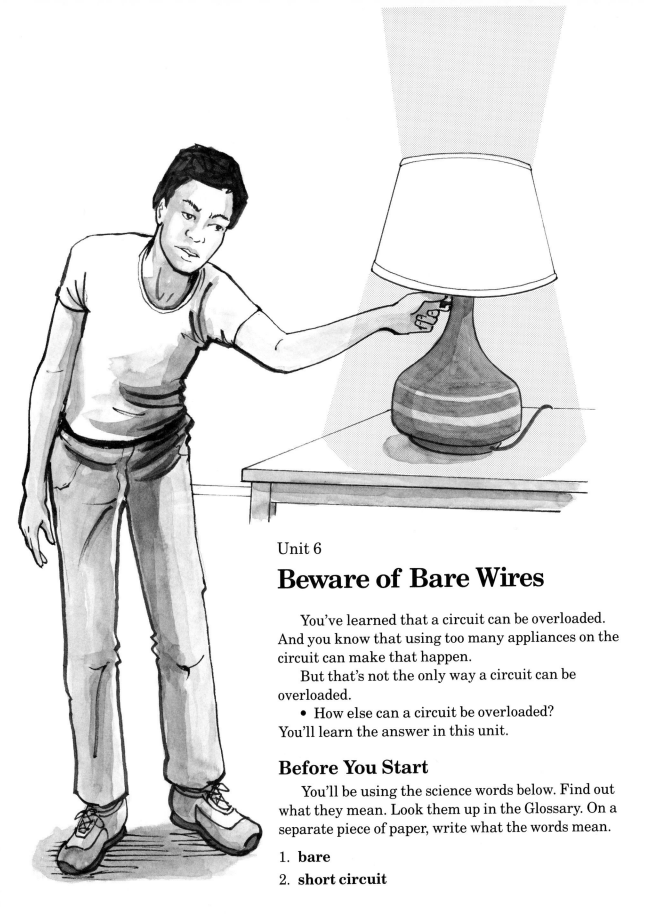

Unit 6

Beware of Bare Wires

You've learned that a circuit can be overloaded. And you know that using too many appliances on the circuit can make that happen.

But that's not the only way a circuit can be overloaded.

• How else can a circuit be overloaded?
You'll learn the answer in this unit.

Before You Start

You'll be using the science words below. Find out what they mean. Look them up in the Glossary. On a separate piece of paper, write what the words mean.

1. **bare**
2. **short circuit**

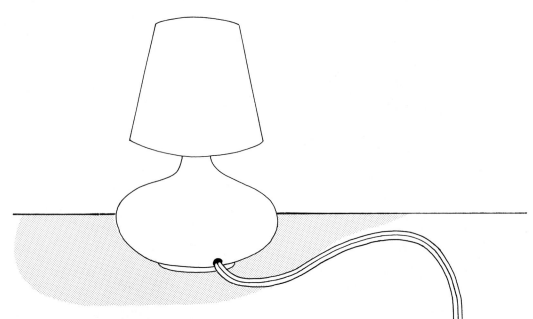

The Shortest Path

The picture on this page shows a lamp plugged into an outlet. There's a place on the cord where the covering has come off. The cord wires are bare, and they're touching each other.

When the lamp is turned on, electricity flows through one cord wire toward the lamp. But it doesn't get to the lamp. Why?

Because it crosses over to the other wire. It crosses over where the two wires touch.

Where does the electricity go after it crosses over to the other wire?

Right! It goes back where it came from.

Electricity always takes the shortest path. That's what it does in the cord. The two bare wires make a short path. They make a short circuit in the cord.

With your finger, trace the path of the electricity as it flows through the short circuit. Start from the outlet. Move your finger up the cord to where the wires cross. Then move your finger back down the cord to the outlet.

You can see that the lamp won't light up when there's a short circuit in the cord. But that's not all that happens. Too much electricity flows in the circuit. The circuit is overloaded. In the next experiment, you'll see what happens then.

Experiment 3

What happens with a short circuit?

Materials

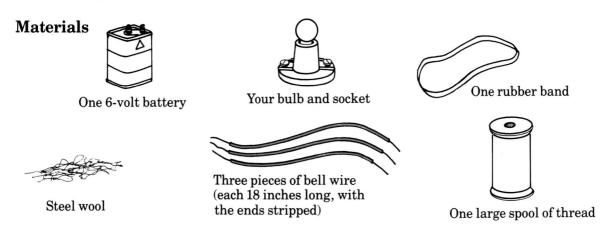

One 6-volt battery

Your bulb and socket

One rubber band

Steel wool

Three pieces of bell wire
(each 18 inches long, with
the ends stripped)

One large spool of thread

Procedure

1. Strip off about one inch of covering
from the middle of a wire. The bare
wire must show.

2. Do the same with another wire.

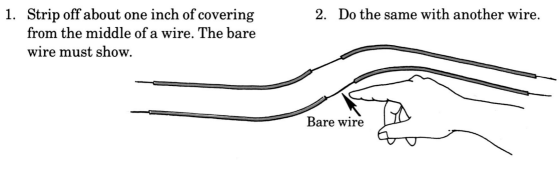

Bare wire

3. Connect the wires to the socket.

4. Pick up a socket wire and the last wire.
Press the wires against the spool.
Make sure that the stripped ends stick
up. (Turn the page for the next step.)

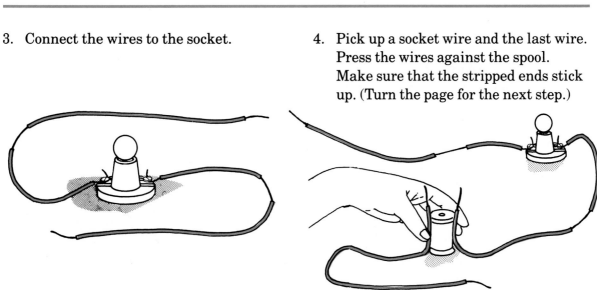

5. Put a rubber band around the wires and the spool. The rubber band will hold the wires in place. Make sure the wires don't touch each other.

6. Take a small piece of steel wool. Stretch it out to make a thin strip. Connect the steel wool to the ends of wire on the spool.

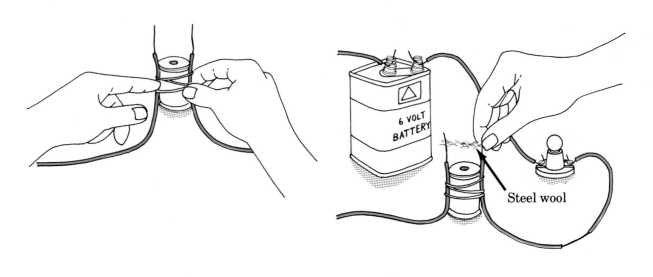

Steel wool

7. Connect the socket wire to a post of the battery. From now on, be careful: Don't let the bare wires touch each other.

8. Connect the wire on the spool to the other post. You've closed the circuit. The bulb should light up.

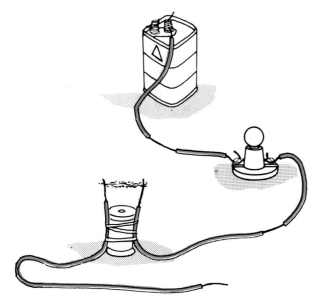

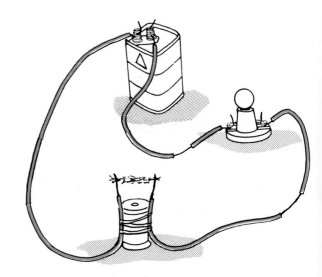

9. Pick up the socket wires. Now you can touch the bare wires together. Do it. Watch the bulb and the steel wool.

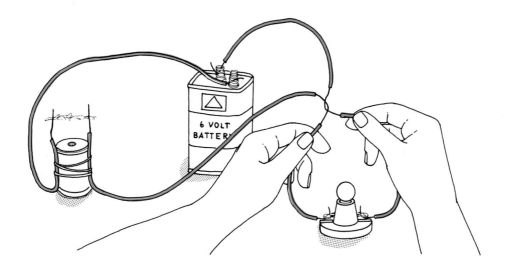

Observations
1. What happens to the bulb when the bare wires touch?
2. What happens to the steel wool when the bare wires touch?

Conclusions
Finish these sentences. Use the words *overloaded, opens, electricity,* and *short.*

1. When the bare wires touch, they make a _____ circuit.
2. Too much _____ flows through the wires.
3. The circuit is _____.
4. The steel wool _____ the circuit like a fuse.

Review

Show what you learned in this unit. Match the words in the list below with their meanings.

covering   hot   short circuit
bare   fuse   overloaded

1. A short path for electricity
2. The kind of wires that make a short circuit
3. What wires become when they're overloaded
4. What keeps a circuit safe
5. Carrying too much electricity
6. The part of a wire that doesn't conduct electricity

Check These Out

1. Check the cords on the appliances you use at home. Find out if the covering is coming off.
2. Find out what to do when an appliance cord has a short circuit in it. Ask someone who works in a hardware store.
3. Here are more things you may want to find out:
   - What is an electric meter? How does it work? How does your local electric company use electric meters?
   - What is electronic music?
   - Why is it dangerous to stand near a tree, a pole, or an open window during a lightning storm?

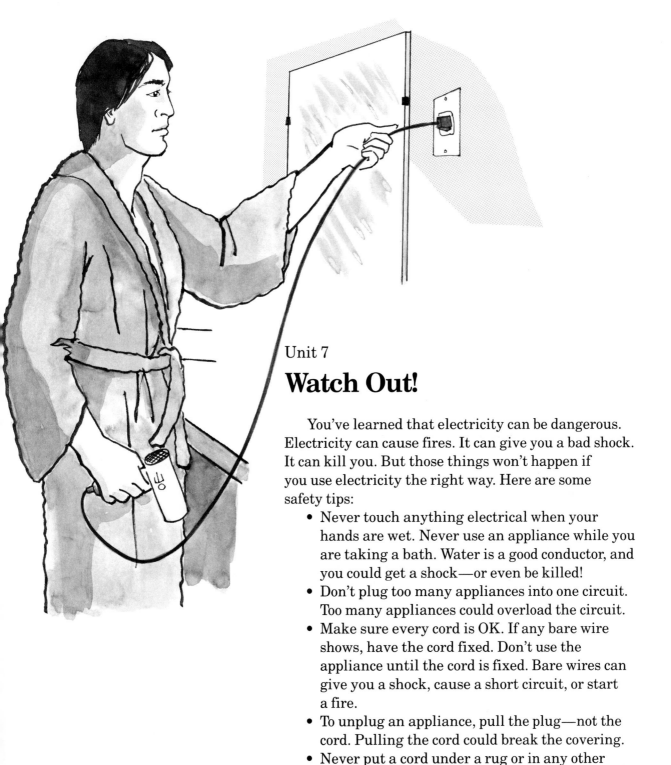

Unit 7

Watch Out!

You've learned that electricity can be dangerous. Electricity can cause fires. It can give you a bad shock. It can kill you. But those things won't happen if you use electricity the right way. Here are some safety tips:

- Never touch anything electrical when your hands are wet. Never use an appliance while you are taking a bath. Water is a good conductor, and you could get a shock—or even be killed!
- Don't plug too many appliances into one circuit. Too many appliances could overload the circuit.
- Make sure every cord is OK. If any bare wire shows, have the cord fixed. Don't use the appliance until the cord is fixed. Bare wires can give you a shock, cause a short circuit, or start a fire.
- To unplug an appliance, pull the plug—not the cord. Pulling the cord could break the covering.
- Never put a cord under a rug or in any other place where people will walk on it. Walking on a cord could break the covering.

What other safety tips can you add to that list?

What's Wrong?

Here are some pictures of people using electricity. But they're using it the wrong way. Tell what they're doing wrong. Use what you have learned. Then check your answers. (The answers are upside down.)

1. What's wrong?
2. What could happen?
3. Why would that happen?
4. How could you keep that from happening?

Answers

1. The covering has come off the cord.
2. The bare wires could touch and blow a fuse.
3. When bare wires touch, they cause a short circuit. Too much electricity flows through the circuit.
4. Have a new cord put on the mixer.

1. What's wrong?
2. What could happen?
3. Why would that happen?
4. How could you keep that from happening?

Answers

1. Too many appliances are plugged into the same outlet.
2. The circuit could become overloaded. It could blow a fuse or start a fire.
3. Too much electricity is being used for one circuit.
4. Plug some of the appliances into another outlet in another part of the room.

1. What's wrong?
2. What could happen?
3. Why would that happen?
4. How could you keep that from happening?

Answers

1. The cord is in a place where people will walk on it.
2. The covering could break and cause a short circuit.
3. Walking on the wire could wear out and break the covering.
4. Put the lamp near the outlet so the cord can be kept out of the way.

1. What's wrong?
2. What could happen?
3. Why would that happen?
4. How could you keep that from happening?

Answers

1. The woman is touching the radio with wet hands. The radio is over the sink.
2. The woman could get a shock.
3. The water on her hands could conduct electricity. If the radio fell into the sink, the water in the sink would conduct electricity.
4. Dry your hands before touching anything that has electricity. Don't put an electric appliance near a sink, bathtub, or other place with water.

Show What You Learned

What's the Answer?

Find the answers for the questions. There may be more than one correct answer to a question.

1. What does electricity flow through?
   a. A closed circuit
   b. An open circuit
   c. A rubber circle
2. What conducts electricity?
   a. Water
   b. Plastic
   c. Metal
3. What does electricity make?
   a. Wire
   b. Heat
   c. Light
4. What kind of circuit is dangerous?
   a. Short
   b. Overloaded
   c. Open

What's the Word?

Give the correct word for each meaning.

1. Special path for electricity
   C _____
2. Something that opens and closes a circuit
   S _____
3. Something that carries electricity
   C _____
4. Thin wire inside a light bulb
   F _____
5. Carrying too much electricity
   O _____
6. Something that keeps a circuit safe
   F _____

Congratulations!

You've learned a lot about electricity. You've learned

- What electricity can flow through
- How we can control electricity
- How we can use electricity safely
- And many other important facts about electricity

SOUND

What kind of force is sound? How are sounds made? How do they travel? What kinds of sounds do we hear? How can we control sounds? In this section, you'll learn many important facts about sounds. And you'll learn how you can make sounds do what you want them to do.

Contents

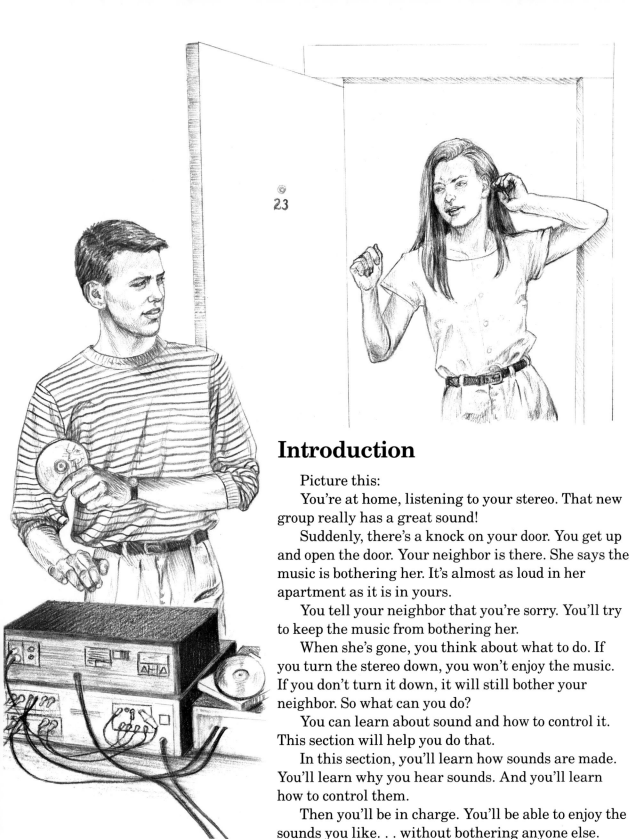

Introduction

Picture this:

You're at home, listening to your stereo. That new group really has a great sound!

Suddenly, there's a knock on your door. You get up and open the door. Your neighbor is there. She says the music is bothering her. It's almost as loud in her apartment as it is in yours.

You tell your neighbor that you're sorry. You'll try to keep the music from bothering her.

When she's gone, you think about what to do. If you turn the stereo down, you won't enjoy the music. If you don't turn it down, it will still bother your neighbor. So what can you do?

You can learn about sound and how to control it. This section will help you do that.

In this section, you'll learn how sounds are made. You'll learn why you hear sounds. And you'll learn how to control them.

Then you'll be in charge. You'll be able to enjoy the sounds you like. . . without bothering anyone else.

Unit 1

Making Sounds

Listen. Do you hear people talking? Cars and trucks going down the street? A telephone ringing? Those are sounds you hear every day.

Sounds are all around you—many different sounds. The sound of your telephone ringing is different from the sound of your friend saying hello. But both of those sounds are made the same way.

• How are sounds made?

You'll learn the answer in this unit.

Before You Start

You'll be using the science words below. Find out what they mean. Look them up in the Glossary that's at the back of this book. On a separate piece of paper, write what the words mean.

1. **motion**
2. **vibration**

Something Has to Move

Suppose you're playing a guitar. You pluck a string. (You pull it and let it go.) You see the string move back and forth. You hear a sound. What do you think makes the sound?

The back-and-forth motion of the string makes the sound. That motion is called vibration.

Every sound is made the same way—by vibration. You can't always see or feel the vibration. But if you hear a sound, something is vibrating.

A fly makes a buzzing sound by vibrating its wings. A telephone rings when something inside the telephone vibrates. How do you think a radio makes sounds?

Right! Something inside the radio vibrates. Put your hand on a radio when the radio is on. You'll feel the vibration.

A fly, a telephone, and a radio make different sounds. But to make any sound, something has to move. No matter what kind of sound you hear, it is made by one thing—vibration.

Feel the Vibes

When you talk or sing, something in your throat is vibrating. You can feel the vibrations in your throat. Here's how:

Put your fingers on the sides of your throat, just below your chin. Now hum. What do you feel?

Do you feel something vibrating? You should.

Put your fingers on the sides of your throat again. Hum again. Then stop. What happens when you stop humming?

Do the vibrations stop? They should.

Here's another kind of vibration you can feel. Put your front teeth together and open your lips. Now make the sound *zzzzzzz*. What do you feel?

You should feel vibrations just behind your teeth.

What happens when you stop making the sound *zzzzzzz*?

Do the vibrations and the sound stop at the same time? They should.

Sounds All Around

You know that whenever you hear a sound, something is vibrating.

Sometimes you can *see* what's vibrating. For example, you can see a guitar string vibrate when you pluck it.

Sometimes you can *feel* what's vibrating. For example, you can feel something vibrate in your throat when you hum.

Things are vibrating all around you. Listen to the sounds those things make. Listen at school, at work, at home, or on the street.

Find things that you can see vibrating. Find things that you can feel vibrating.

Keep a record of what you find. Make a list of the things you find that you can see vibrating. Make a list of things that you can feel vibrating. Some things may go on both lists.

Experiment 1
What vibrates to make the sound?

Sometimes when you hear a sound, it's easy to tell what's vibrating. Sometimes it's not so easy.

In this experiment, you'll make two different sounds. And you'll figure out what vibrates to make each sound.

Materials (What you need)

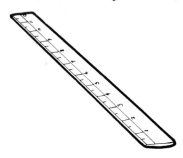

One thin 12-inch ruler

Three heavy books

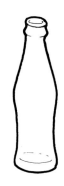

One empty soda bottle

Procedure (What you do)

1. Make the first sound: Stack the books at the edge of a table. Place the ruler under the stack of books. Let about 8 inches of the ruler stick out beyond the edge of the table.

2. Now push the end of the ruler down and quickly let it go. Do this two or three times. Listen each time.

3. Make the second sound: Get the bottle. Hold it so that the top of the bottle touches your bottom lip.

4. Now, with your lips almost closed, blow across the top of the bottle. Do this two or three times. Listen each time.

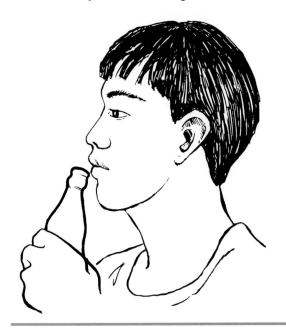

Observations (What you hear)

1. What happens each time you push the ruler down? Do you hear a sound? You should.
2. What happens each time you blow across the bottle? Do you hear a sound? You should.

Conclusions (What you learn)

Choose the correct answer to each question.

1. What vibrates to make the first sound?
   a. The books
   b. The ruler
   c. Your hand
2. What vibrates to make the second sound?
   a. Your lips
   b. Something in your throat
   c. Air in the bottle

Review

Show what you learned in this unit. Finish the sentences. Match the words on the left with the correct words on the right.

1. Vibration is
2. Whenever you hear a sound,
3. When the vibration stops,
4. You can see vibration
5. You can feel vibration in your throat

a. the sound stops.
b. back-and-forth motion.
c. when you talk or sing.
d. something is vibrating.
e. when you pluck a guitar string.

Check These Out

1. Make a Science Notebook for Sound. Put your list of glossary words and their meanings in the notebook. Also keep a record of the experiments you do. You can put anything else you learn about sound in your Science Notebook.
2. Get these things: a sheet of paper, a bottle, a ruler, a rubber band, and some coins. Use them to make as many different sounds as you can.
3. Find some musical instruments at your school. Play them, or ask someone else to. Figure out what vibrates in each instrument.
4. As you work through this section, you may want to find out more about sound. You can find out more by looking in a dictionary or an encyclopedia, or by getting books about sound from a library. You can also talk with an expert, such as a sound engineer, a musician, or a physics teacher.

   Here are some things you may want to find out:
   - What is a tuning fork? How does it work?
   - What are vocal cords? How do they make sounds?
   - Some insects vibrate parts of their bodies to make sounds. How does a grasshopper make sounds?

Unit 2

High Sounds and Low Sounds

What's the difference between a woman's voice and a man's voice? Usually a woman's voice is higher and a man's voice is lower.

You know that sounds, such as voices, are made by vibrations. But vibrations aren't all the same. So sounds aren't all the same either.

Some sounds are high, like the sound of a siren on a fire engine. Other sounds are low, like the sound of a dog growling.

- Why are some sounds high?
- Why are some sounds low?

You'll learn the answers in this unit.

Before You Start

You'll be using the science words below. Find out what they mean. Look them up in the Glossary. On a separate piece of paper, write what they mean.

1. **frequency**
2. **pitch**

Different Speeds

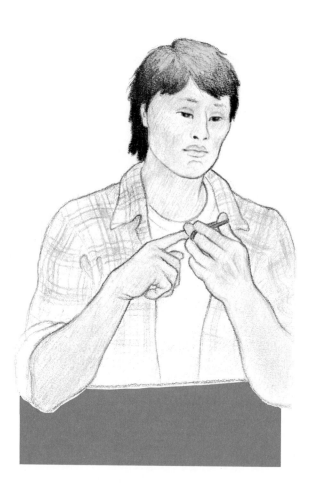

Get a rubber band. Stretch it a little. Pluck it. Notice how fast it vibrates.

Now stretch the rubber band more. Pluck it again. Does it vibrate faster or slower than before?

Right! It vibrates faster.

A thing can vibrate at different speeds. It can vibrate fast, or it can vibrate slowly.

You learned that when a thing vibrates, it makes a sound. The sound can be high or low. That depends on how fast the vibrations are.

Get a rubber band again. Stretch it a little. Make it vibrate. Listen to the sound it makes. Then stretch the rubber band more. Make it vibrate faster. Listen to that sound. Is that sound higher or lower than the first sound?

Right! That sound is higher. It's higher because the rubber band is vibrating faster.

You've seen that a thing can vibrate at different speeds. There's a name for the speed of vibration. The speed of vibration is called frequency.

If a thing vibrates fast, it has a high frequency. And it makes a high sound.

If a thing vibrates slowly, what kind of frequency does it have?

What kind of sound does it make?

That's right! It has a low frequency, and it makes a low sound. You'll find out more about high and low sounds on the next page.

Hearing the Pitch

People use the word *pitch* to talk about sounds. The pitch of a sound is how high or low the sound is. A high sound has a high pitch. A low sound has a low pitch.

Here are three things that make sounds with a high pitch: a siren, a whistle, and a bird.

Think of two more things that make sounds that have a high pitch.

Here are three things that make sounds with a low pitch: a cow, a tuba, and a foghorn.

Think of two more things that make sounds that have a low pitch.

Now make some sounds. Use your voice or things in the room. Keep a record of which sounds have a high pitch and which ones have a low pitch.

Changing Pitch

Imagine going to a concert where all the instruments in the band could make sounds with only one pitch. It would be pretty boring! Music means putting together sounds with different pitches. So instruments must be able to change the pitch of the sounds they make.

You learned that a low pitch is made by something vibrating slowly. You also learned that a high pitch is made by something vibrating fast. So if you change how fast something vibrates, the sound it makes will change in pitch.

Remember what you heard when you plucked the rubber band. The tighter you stretched it, the higher the pitch. So stretching something is one way of changing the pitch of its sound.

What is another way of changing the pitch of the sound something makes?

Do this experiment and find out.

Experiment 2

How can pitch be changed?

Materials

One thin 12-inch ruler   Three heavy books

Procedure

1. Set up the ruler and books as you did in Experiment 1. Let 9 inches of the ruler stick out over the edge of the table.

2. Now flick the end of the ruler as you did before. Listen to the pitch of the sound. You can get a louder sound by having a friend push down on the stack of books.

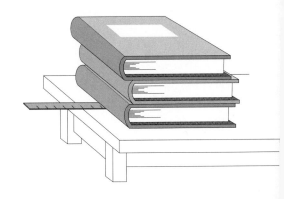

3. Move the ruler so that 7 inches of it stick out over the edge. Flick the ruler again. Listen to the pitch.

4. Try flicking the ruler with 5 inches sticking out, and then 3 inches. Pay attention to the pitch each time.

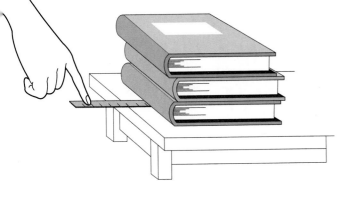
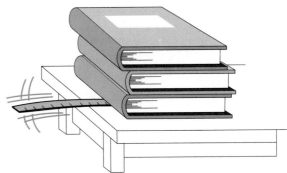

Observations

1. How did the pitch in step 2 compare to the pitch in step 3?
2. How much of the ruler was sticking out when you made the sound with the highest pitch?

Conclusions

Choose the correct way to finish each sentence.

1. The pitch of the vibrating ruler depends on
   _____.
   a. the length of the vibrating part
   b. how hard you flick the ruler
   c. how heavy the books are
2. The shorter the length of the vibrating part of the ruler, the _____ it vibrates.
   a. slower
   b. faster
   c. softer
3. The faster the ruler vibrates, the _____ its frequency.
   a. lower
   b. higher
   c. louder
4. The higher the frequency of the vibrations, the _____ the pitch of the sound.
   a. lower
   b. higher
   c. softer

Make a Musical Instrument

A musical instrument makes sounds of different pitches. Think about a guitar. Each string on a guitar has a different pitch.

You can make a musical instrument that has five different pitches. You'll need these materials:

- Five glasses, all the same size
- One large pitcher of water
- One metal spoon

Here's what to do:

---

1 Fill one glass almost to the top with water.

2 Pour water into another glass. Make it a little less full than the first glass.

---

3 Pour water into the other three glasses. Pour a different amount into each glass.

4 Tap the glasses with the spoon. Listen as you tap each glass. You should hear five different pitches.

5 Pick up the glass that has the lowest pitch. Put it at your left. Then line up the other four glasses in order, from lowest pitch to highest pitch.

You've made a musical instrument! Play a tune on it by tapping the glasses with the spoon.

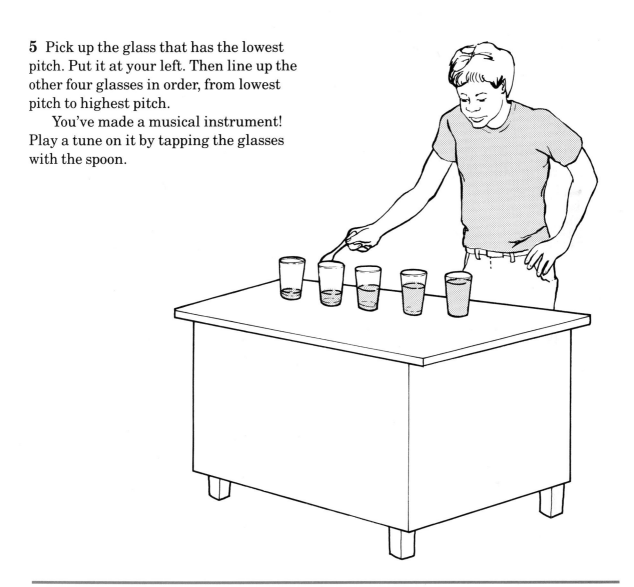

What's Happening?

Use the word *highest* or *lowest* to finish each sentence.

1. The glass with the most water has the _____ pitch.
2. The glass with the least water has the _____ pitch.

Review

Part I

Show what you learned in this unit. Match the words in the list below with their meanings.

vibrate   frequency   pitch

1. Speed of vibration
2. Move back and forth
3. How high or low a sound is

Part II

Answer these questions about pitch.

4. Which has a higher pitch: a record played at 45 or at 33?
5. Which has a lower pitch: tapping on a glass full of water or on one that's almost empty?

Check These Out

1. Look inside a piano. Notice the strings. Some strings are longer than others. When the piano is played, the strings vibrate. Which strings make low sounds? Which strings make high sounds?

2. Find pictures of musical instruments in magazines and newspapers. Cut out the pictures. Put them in three groups: instruments you pluck, instruments you blow into, and instruments you hit. For each instrument, figure out what vibrates.

3. Make a "box guitar." Get an empty tissue box with a large opening. Get a few rubber bands—some thick ones and some thin ones. Put the rubber bands around the box the long way. Play the guitar by plucking the rubber bands over the opening.

4. Here are more things you may want to find out:
   - What is ultrasonic sound? How do doctors use it? How does the U.S. Navy use it?
   - Why does each string on a guitar have a different pitch?
   - Some animals can hear sounds that are too high or too low for people to hear. Which animals can do that?

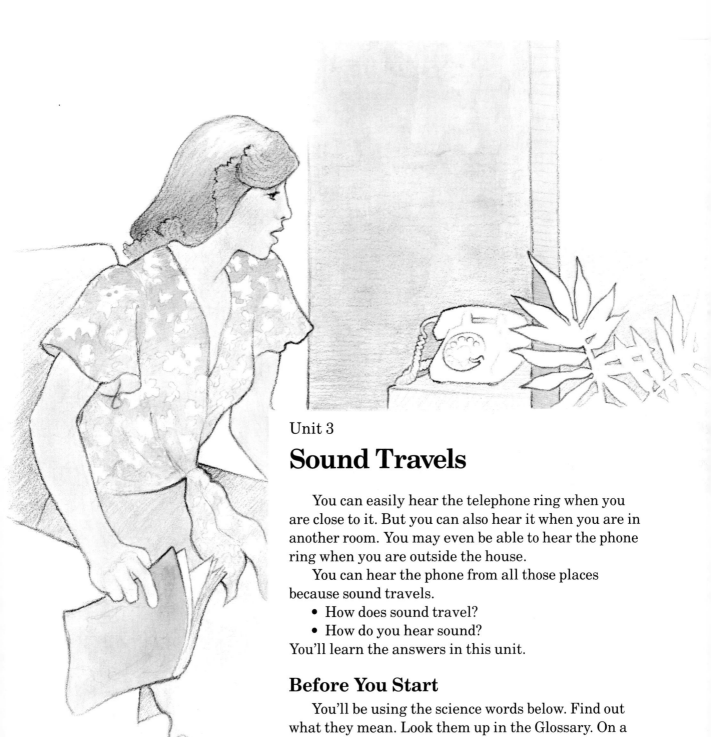

Unit 3

Sound Travels

You can easily hear the telephone ring when you are close to it. But you can also hear it when you are in another room. You may even be able to hear the phone ring when you are outside the house.

You can hear the phone from all those places because sound travels.

- How does sound travel?
- How do you hear sound?

You'll learn the answers in this unit.

Before You Start

You'll be using the science words below. Find out what they mean. Look them up in the Glossary. On a separate piece of paper, write what the words mean.

1. **object**
2. **surrounding**

Traveling Through Air

Think of an alarm clock going off. You know that the sound of the alarm starts with vibrations. For you to hear that sound, the vibrations have to travel. They have to travel from the clock to your ears.

What do you think the sound travels through?

The sound travels through the air. It travels through the air that's between the clock and your ears.

How does sound travel through air? This is what happens:

An object vibrates—it moves back and forth. That vibration makes the air right next to the object vibrate. (Look at picture **A**.)

That vibrating air makes air farther from the object vibrate. (Look at picture **B**.)

What happens then? (Look at picture **C**.)

Right! Air farther and farther from the object vibrates.

So a vibrating object sets off vibrations in the surrounding air. Those vibrations travel through the air to your ears. And that's when you hear the sound.

Listen for a minute. What are three sounds that are traveling through the air right now?

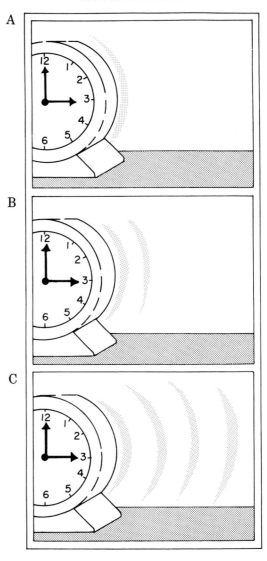

How sound travels

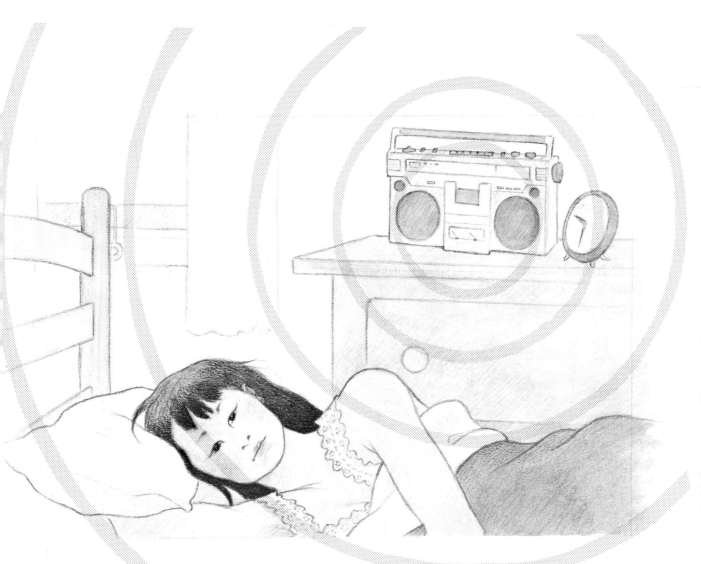

Up, Down, and All Around

Suppose you're lying in bed, listening to the radio. You're right in front of the radio. Then you get up and move away from the radio. Where else in the room can you hear it from?

You can hear the radio from anywhere in the room.

You learned that a vibrating object sets off vibrations in the surrounding air. Those vibrations are called **sound waves**.

Sound waves move away from the vibrating object. They spread out in all directions. They travel up, down, and all around.

That's why you can hear the radio from anywhere in the room. You can hear it when it's in front of you or behind you. And you can also hear it when it's above you or below you.

Suppose sound didn't travel in all directions. Suppose it traveled only in a straight line. Could you still hear the radio from anywhere in the room?

No, you couldn't. Where would you have to be to hear the radio?

How You Hear

How do you hear the sound of a radio? The picture on this page shows you how. Read the steps.

1. A special part in the radio vibrates. That vibration sets off sound waves.

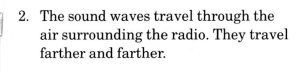

2. The sound waves travel through the air surrounding the radio. They travel farther and farther.

3. The sound waves enter your ears.

4. When the waves reach certain parts inside your ears, those parts vibrate. And that's when you hear the sound.

What's the Order?

The steps below tell you how you hear the sound of a radio. But the steps are out of order. On a separate piece of paper, list the steps in the right order. Number each step.

- The sound waves travel through the air surrounding the radio.
- The sound waves enter your ears.
- A part in the radio vibrates and sets off sound waves.
- Certain parts inside your ears vibrate, and you hear the sound.

Sounding Off

You can see for yourself that sound waves travel in all directions. Here's how:

Get a record player. Put it on a table in the middle of the room. Play a record. Now go to each place listed below. Can you hear the music from that place? Keep a record of the places where you can hear the music and where you can't hear it.

1. The front of the room
2. The back of the room
3. Close to the floor
4. Up on a chair
5. By the door
6. Next to a window

You listened from six different places. From how many of those places could you hear the music?

You could probably hear the music from all six places. Why?

Review

Show what you learned in this unit. Finish the sentences. The words you'll need are listed below. One word will be left over.

waves   directions   object
alarm   surrounding   vibrate

1. All sounds start from a vibrating _____.
2. Sound _____ are vibrations in the air.
3. A vibrating object sets off vibrations in the _____ air.
4. Sound waves spread out in all _____.
5. Sound waves make certain parts in your ear _____.

Check These Out

1. Find out more about how people hear. Ask a doctor who takes care of hearing problems to visit your class. Before the doctor visits, write down some questions to ask him or her.
2. Some people who can't hear use sign language to "talk" to other people. Find out about sign language. Learn how to sign your name.
3. Some deaf people can dance to music. Find out how they "hear" the music.
4. Here are more things you may want to find out:
   - How do hearing aids work?
   - What is an oscilloscope? What is it used for?
   - What is the Doppler effect? What causes it?
   - What are molecules? What happens to molecules when sound travels?
   - What happens to sounds outdoors on a windy day?

Unit 4

What Sound Travels Through

Suppose you're driving along in your car. All the windows are closed. Suddenly you hear the siren of a fire truck. You hear it clearly even though the windows are closed. Why? Because sound travels. It travels through different kinds of things.

• What kinds of things can sound travel through? You'll learn the answer in this unit.

Before You Start

You'll be using the science words below. Find out what they mean. Look them up in the Glossary. On a separate piece of paper, write what the words mean.

1. **gas**
2. **liquid**
3. **solid**

Traveling Sounds

The bell inside a telephone vibrates and sets off sound waves. For you to hear the sound, those sound waves must travel. They must travel from the vibrating bell to your ears. And they must travel through something. What do the sound waves travel through?

Right! The sound waves travel through air.

Most sounds that you hear travel through air. That's because your ears are usually surrounded by air. And so are most of the objects that make sounds.

But sounds travel through other things besides air, which is a gas. They travel through solids, such as wood or metal. They also travel through liquids, such as water.

Because sound travels through many kinds of things, you can hear sounds from far away. You can even hear sounds through walls.

Imagine you are indoors. You hear a jet plane flying high over your head. What do you think the sound travels through to reach your ears?

Right! The sound travels through air.

Suppose you are in a closed room. You hear the sound of a TV from another room. What does the sound travel through to reach your ears?

The sound travels through air. It also travels through the walls of the room.

Every sound you hear travels through something. Otherwise you would not be able to hear it.

Sound Through String

You know that sound travels through air. Air is a gas.

Can sound also travel through a solid? You and a partner can find out. You will need these things:

- Two metal spoons
- One piece of string, about 3 feet long

1 Tie one spoon to the middle of the string. Hold the ends of the string so that the spoon hangs down.

2 Have your partner tap the hanging spoon with the other spoon. Listen. You'll hear a sound. What does that sound travel through?

Right! That sound travels through air, which is a gas.

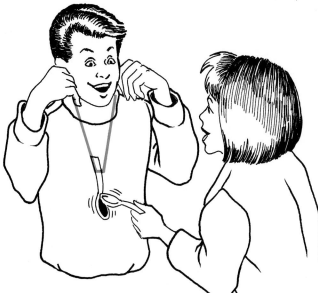

3 Now put the ends of the string in your ears. Have your partner tap the hanging spoon again. Listen again. You'll hear another sound. What does that sound travel through?

Right! That sound travels through the string. The string is made of a solid. On the next page, you'll learn more about some other solids that sound can travel through.

Through Solids

You learned that sound can travel through a string. Can sound also travel through other solids? Find out for yourself. Here's how:

Get a partner. Stand at one end of a table. Ask your partner to stand at the other end. Have your partner tap the table. You'll hear a sound that travels through air.

Now put your ear against the edge of the table. Cover your other ear with your hand. Have your partner tap the table again. Do you hear a sound?

What does the sound travel through?

Right! The sound travels through the table. The table is made of a solid. What solid is the table made of?

The table is probably made of wood. But it may be made of metal or plastic. Wood, metal, and plastic are all solids.

Next do this: Go outside the room. Leave the door open. Ask your partner to stay inside the room and call your name. You'll hear your partner's voice. It travels through air.

Now close the door. Put your ear against the door. Have your partner call your name again. Do you hear your partner's voice?

What does the voice travel through?

Right! Your partner's voice travels through the door. The door is made of a solid. What solid is the door made of?

Through Liquids

You found out that sound can travel through a table or a door. Look around you. Find three more things that sound can travel through. Find things that are made of solids, such as wood, metal, plastic, or glass.

You learned that sound travels through gases and solids. What else can sound travel through?

Yes! Sound can travel through liquids.

To make sure of this, work with your partner again. Get two coins and a big glass bowl of water.

Have your partner tap the two coins together. You'll hear a sound that travels through air.

Now put your ear against the bowl. Have your partner hold the coins under water and tap them together again. You'll hear a sound. What does the sound travel through?

Right! The sound travels through water (and, of course, the bowl). Water is a liquid.

Making a String Telephone

You can make a string telephone. You'll need these materials:

- One pencil
- Two paper cups
- One piece of string (about 8 feet long)
- Two paper clips

Get a partner and do this:

---

1 With the pencil, poke a small hole in the bottom of each cup.

Hole

2 From the outside of each cup, push one end of the string through the hole. Push about 6 inches of string through.

---

3 Tie each end of the string to a paper clip. (The clips will hold the string in place.)

Paper clip

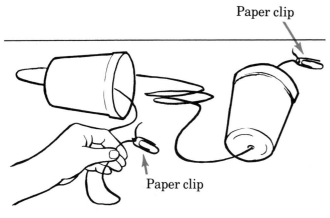

Paper clip

4 Give one cup to your partner. Keep the other cup. Move away from your partner until the string is tight. Be sure the string isn't touching anything.

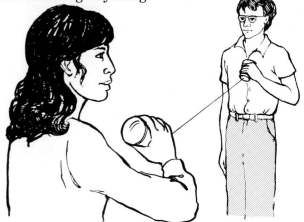

5 Hold your cup to your ear. Tell your partner to speak softly into the other cup. Have your partner ask a question. Can you hear the question? If you can, answer it.

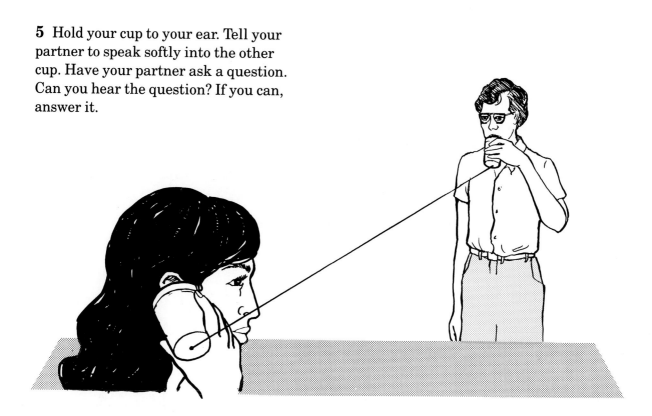

What's Happening?

1. Your partner spoke softly into the cup. But you could hear your partner's voice. Why?

    You could hear your partner's voice because the sound made the string vibrate. The sound traveled through the string.

2. Your string telephone won't work unless the string is pulled tight. Why?

    The string won't vibrate unless it's pulled tight.

3. Your string telephone won't work if the string touches anything. Why?

    If the string touches anything, the vibrations will stop.

Review

Show what you learned in this unit. Finish the sentences. The words you'll need are listed below.

vibrates   waves   travels
sounds   gas   solids
water

1. Sound waves are set off when something
   v_____.
2. For you to hear, sound w_____ must travel to your ears.
3. Every sound t_____ through something.
4. Most s_____ you hear travel through air.
5. Air is a g_____.
6. Sound waves can travel through s_____, such as metal or wood.
7. Sound waves can also travel through liquids, such as w_____.

Check These Out

1. Find out how a telephone works. Ask someone from a telephone company to explain it to your class.
2. Sounds travel faster through the ground than they do through the air. Hunters know that. So they sometimes put an ear to the ground to listen for animals. What are other ways that people listen to sounds through solids? Draw a picture that shows one of those ways.
3. Here are more things you may want to find out:
   * What is the sound barrier? What happens if an airplane goes faster than the speed of sound?
   * Why do you hear thunder after you see the lightning that causes it?
   * Whales and porpoises can send sounds that travel 100 miles. How do they do that?

Unit 5

Sound Changes Direction

Picture this:

A school bell rings. The air surrounding the bell vibrates. And sound waves travel out in all directions.

Some of the sound waves hit walls. And some of the waves hit curtains. Something happens to those waves.

- What happens when sound waves hit hard objects?
- What happens when sound waves hit soft objects?

You'll learn the answers in this unit.

Before You Start

You'll be using the science words below. Find out what they mean. Look them up in the Glossary. On a separate piece of paper, write what the words mean.

1. **absorb**
2. **reflect**
3. **surface**

Sound Bounces

Sound waves travel away from a vibrating object. They keep traveling away from the object unless they hit a hard surface. What do you think will happen when the sound waves hit a hard surface?

The sound waves will bounce. The hard surface will reflect most of the sound waves.

Picture this:

You're playing drums in an empty room. You hit a drum. Some sound waves travel toward the wall. What happens when they hit the wall?

The wall reflects most of the sound waves. The sound waves bounce off the wall.

Suppose the sound waves bounce all the way across the room. They hit another wall. What happens then?

Right! That wall reflects the sound waves again.

The sound waves may bounce several times. Finally, they will slow down and stop.

Sound waves bounce off hard surfaces. So you sometimes hear the same sound more than once. You hear the sound when it first happens. Then you hear it again when a hard surface reflects the sound waves back to your ears.

What do you call a sound that bounces back to your ears?

You call that sound an **echo**.

You might hear echoes in a large cave. Why?

Right! The walls of the cave reflect the sound waves.

Sounding Louder

Suppose you shout. A hard surface reflects the sound. You might hear an echo.

But if you're too close to the hard surface, you won't hear the echo. The echo will reach your ears very quickly. You will hear the sound and the echo at the same time. The echo will make the sound louder.

Try making a sound louder. You'll need an empty container. It should be tall and round, like an oatmeal box.

1 Say, "Ha, ha, ha." Listen to the sound.

2 Hold the container up to your mouth. Say, "Ha, ha, ha" into the container. Listen to the sound.

What's Happening?

1. How is the sound different when you use the container? Is the sound louder? It should be.
2. Why is the sound louder?

   Right! The surfaces of the container reflect the sound. You hear the sound and the echo at the same time.

Sending and Collecting Sounds

Suppose your friend is standing across the street. You shout, but your friend can't hear you. Then you cup your hands around your mouth and shout again. What happens?

Right! Your friend hears you. Your hands keep the sound waves from spreading out. That way you send all the sound waves in one direction—toward your friend.

You can send sound waves in any direction. But you need something like your cupped hand to keep the waves from spreading out.

Now suppose your friend shouts back to you. You can't hear him.

So you cup your hands behind your ears. What happens?

You can hear your friend. Your hands collect the sound waves.

Even without your hands, your ears are shaped to collect sound waves. But with your hands behind them, your ears collect more sound waves.

Try it. Cup your hands behind your ears. Slowly turn your head from side to side. Listen. You should hear more. Why?

Right! Your hands help your ears collect sound waves.

Making Sounds Quieter

You've learned that a hard surface, such as a wall, reflects sound waves. It makes sounds louder. But a soft surface, such as a curtain, doesn't reflect sound waves very well.

A soft surface has many tiny holes in it. When sound waves hit the soft surface, some of the waves enter those holes. The soft surface absorbs sound waves. It makes sounds quieter.

So you can use soft things to make sounds quieter. What are some soft things that can absorb sound waves?

Pillows, rugs, or anything else that's made of cloth can absorb sound waves.

Suppose you play music in an empty room. The hard walls and floors will reflect the sound waves and make the music louder. Now suppose you put rugs, pillows, and curtains in the room. What happens to the music?

Right! The music sounds quieter. Why?

The music sounds quieter because the soft things absorb sound waves.

Experiment 3

What can keep a hard surface from reflecting sound waves?

Materials

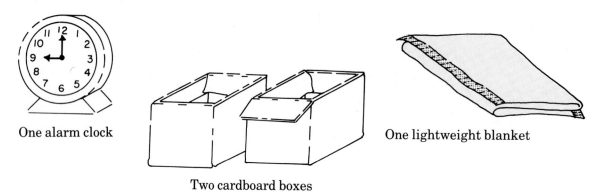

One alarm clock

Two cardboard boxes

One lightweight blanket

Procedure

1. Put the boxes on a table. Line one box with a blanket.

2. Get the alarm clock. Make the alarm go off. While the alarm is ringing, put the clock in the empty box. Listen to the sound.

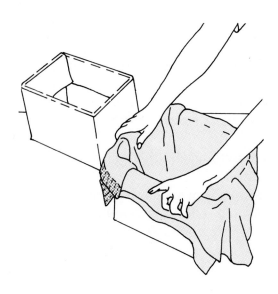

3. Let the alarm go on ringing. Move the clock to the lined box. Listen to the sound now. Is it louder or quieter than before?

4. Move the clock from one box to the other several times. Listen and compare the sounds.

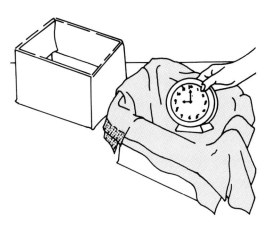 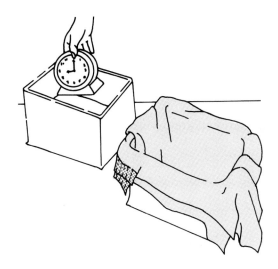

Observations

The empty box has a hard surface. The lined box has a soft surface.

1. Does the sound get louder or quieter when you move the clock from the empty box to the lined box?
2. Does the sound get louder or quieter when you move the clock back to the empty box?

Conclusions

A hard surface reflects sound waves. It makes sound waves louder.

What can keep a hard surface from reflecting sound waves?

a. Putting something hard on the hard surface
b. Putting something soft on the hard surface
c. Putting a clock on it

Review

Show what you learned in this unit. Finish the sentences. The words you'll need are listed below.

echo	spreading	louder
absorb	reflect	quieter

1. Hard surfaces _____ sound waves and make sounds louder.
2. A sound that bounces back to your ears is called an _____ .
3. If you keep sound waves from _____ , a sound will be louder.
4. Soft surfaces _____ sound waves and make sounds quieter.
5. A sound is _____ when you hear it and its echo at the same time.
6. You can use soft things to make a sound _____ .

Check These Out

1. Find out what a megaphone is. Make one out of newspaper or a large paper bag.
2. Some people work with machines that make loud sounds. Those people often develop hearing problems. Find a story about this in a newspaper or a magazine. Tell the class what's being done about it.
3. Visit a concert hall. Find out how the different surfaces in the hall make the music sound better. Tell the class what you learned.
4. Here are more things you may want to find out:
   - What is a decibel? Why do scientists measure decibels?
   - What is a whispering gallery? How does it work?
   - What is sonar? How do bats use it?
   - When a concert hall is full of people, music sounds different from the way it sounds when the hall is empty. How does it sound different? Why does it sound different?

Unit 6

Using What You Know

Suppose you want to listen to music. But someone in your house wants to take a nap. If the music is too loud, it will bother the other person. But if the music is too quiet, you won't be able to hear it. Can you do anything to control the sound? You can.

- What is the science of sound control?
- How can you make a stereo sound louder without turning up the power?
- How can you keep sound in a room?

You'll learn the answers in this unit.

Before You Start

You'll be using the science words below. Find out what they mean. Look them up in the Glossary. On a separate piece of paper, write what the words mean.

1. **acoustics**
2. **amplify**
3. **soundproof**

Controlling Sound

Picture this:

Your stereo is in the living room. You want it to sound louder so that you can hear it in the bedroom. What can you do?

You can turn up the power. But if you turn up the power, the stereo might bother your neighbors.

So what else can you do? How can you amplify the sound of the stereo without turning up the power? You can use what you know about acoustics. Acoustics is the science of sound control.

You've learned a lot about acoustics in this section. You know how sound travels. You know what happens when sound waves hit hard surfaces or soft surfaces.

People who build buildings use acoustics. They build some places to amplify sounds—to make sounds louder.

A concert hall amplifies sounds. What do you think a builder can put in a concert hall to amplify sounds?

To amplify sounds, a builder can put in concrete walls, a wooden stage, and metal chairs. Those things have hard surfaces. They reflect sound waves and make sounds louder.

Builders also use acoustics to soundproof some places. They want to keep sounds from going in or out.

Most movie theaters are soundproofed. What do you think a builder can put in a movie theater to soundproof it?

A builder can put carpets on the floor, curtains on the walls, and cloth on the seats. Those things all have soft surfaces. They absorb sound waves and keep sounds from going out of the theater.

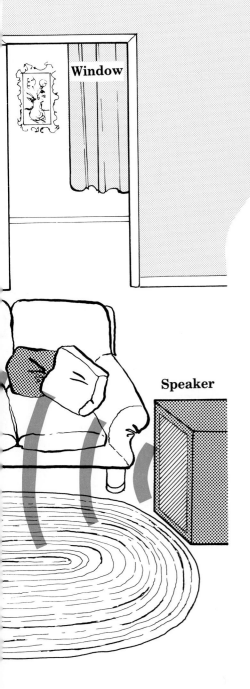

Window

Speaker

Making Sound Louder

Look at the living room and bedroom in the picture. Suppose they're part of your home. Your stereo is in the living room. You want to hear it in the bedroom without bothering your neighbors. Use what you know about acoustics to do that.

1. Look at the speakers. Each speaker sends sound waves straight ahead. What can you do with the speakers to make the sound louder in the bedroom?

   Right! You can turn the speakers so that they face the doorway of the bedroom. Why would that make the sound louder in the bedroom?

2. Notice the large sofa between the speakers and the doorway. The soft surface of the sofa absorbs sound waves. What can you do with the sofa to make the sound louder in the bedroom?

   Right! You can move the sofa away from the speakers. Why would that make the sound louder in the bedroom?

3. Look at the window in the bedroom. It's covered with heavy curtains. What happens when sound waves hit the curtains?

   Right! The curtains absorb sound waves. What can you do to the curtains to make the sound louder in the bedroom?

   Right! You can open the curtains. Why would that make the sound louder in the bedroom?

Soundproofing

Let's say you like to watch TV in your bedroom. But other people who live in your house don't want to listen to the sound. What can you do?

You can close the bedroom door. But people in other rooms can still hear the sound.

What else can you do? How can you keep the sound of the TV from going out of the bedroom? You can soundproof your bedroom.

Builders use what they know about acoustics to soundproof rooms. Here are some things they do to walls and ceilings:

- They build thick double walls. The walls are filled with something soft that absorbs sound waves.
- They put special **acoustical tile** on walls and ceilings. The tile has many tiny holes that absorb sound waves.

Builders do other things to soundproof floors and windows. What do you think builders do to soundproof floors? (Hint: They use something soft.)

What do you think builders do to soundproof windows? (Hint: They use something soft.)

Making Sound Quieter

Look at the bedroom in the picture. Suppose it's your bedroom. Your TV is in the bedroom. You want to be able to hear it. But you don't want to bother anyone. Use what you know about acoustics to do that.

1. What hard surfaces do you see in the room?
2. What happens when sound waves hit those hard surfaces?
3. To soundproof the room, what changes can you make?

Here are some changes you can make:
- Put a rug on the floor.
- Put curtains on the windows.
- Put pillows on the chair and the bed.
- Put cloth or acoustical tile on the walls and the ceiling.
- Move the TV so that it doesn't face the door.

You've shown that you know a lot about acoustics. You know how to amplify sounds. And you know how to soundproof a room. Use what you know to control sound at home and at school.

Show What You Learned

What's the Answer?

Find the correct answer for each question.

1. What is every sound made by?
   a. A musical instrument
   b. A vibrating object
   c. A rubber band

2. What gives a sound a high pitch?
   a. Loud voices
   b. Slow vibrations
   c. Fast vibrations

3. How do sounds travel?
   a. In waves
   b. In groups
   c. In a line

4. What must a sound do for you to hear it?
   a. Bounce off a wall
   b. Travel through something
   c. Have a high pitch

5. What does a hard surface do to sound waves?
   a. Absorbs them
   b. Soundproofs them
   c. Reflects them

What's the Word?

Give the correct word for each meaning.

1. Back-and-forth motion
   V _____

2. How fast something vibrates
   F _____

3. How high or low a sound is
   P _____

4. Throw back sound waves
   R _____

5. Soak up sound waves
   A _____

6. Make louder
   A _____

7. The science of sound control
   A _____

Congratulations!

You've learned a lot about sound. You've learned

- How sounds are made
- How sound travels
- How to control sound
- And many other important facts about sound

MACHINES

How do machines increase our force? How do machines help us do work? What are machines made of? What are the five simple machines? How do they work? In this section, you'll learn many facts about machines. And you'll learn how we use machines to make our lives easier.

Contents

Introduction

Remember *Superman*? Maybe you've read about him or seen him in the movies. He's very strong. He can lift several people at the same time. He can lift buildings, cars, and heavy rocks. He can cut through metal. He can also move faster than anyone else.

Superman came out of somebody's mind. Someone thought him up and wrote stories about him. But there *are* people like Superman in real life. *You* could be one of them. *You* can lift people, buildings, cars, and heavy rocks. *You* can cut through metal. And *you* can move faster than anyone can run.

How can you do all those things? By using **machines**! That's right. Machines can make you stronger. And they can make you move very fast.

In this section, you'll learn about different machines. You'll learn what they can do for you, and how they make you stronger.

You'll also learn about some machines to use for different kinds of work. Someday, you'll have a job. Maybe you already do. This section can show you how to make your work easier!

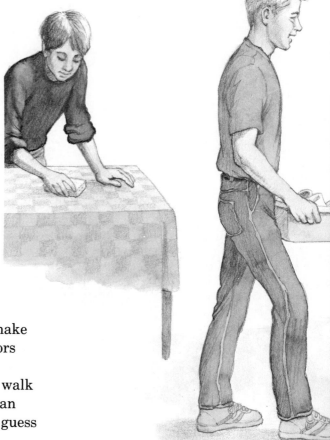

Unit 1

Making Work Easy

It's Saturday! You're working at home. You make your bed and wash the dishes. You sweep the floors and scrub the tub. Finally your work is done.

"No more work today!" you tell yourself. You walk to a friend's house. You and he will take a bus to an amusement park and ride the roller coaster. But guess what: You're still doing work!

• What is work?
• What things make your work easier?

You'll learn the answers in this unit.

Before You Start

You'll be using the science words below. Find out what they mean. Look them up in the Glossary that's at the back of this book. On a separate piece of paper, write what the words mean.

1. **force**
2. **tool**
3. **work**

Doing Work

Are you doing work when you sweep the floor? when you're playing soccer? when you're thinking about how to solve a math problem?

You probably answered that both sweeping the floor and thinking about a math problem are work. You probably said playing soccer is not work.

But a scientist would answer differently. To a scientist, sweeping the floor and playing soccer are work and thinking is not. That is because scientists have a special meaning for the word *work*.

Scientists say that **work** is moving something from one place to another.

What do you move when you play soccer? Right—you move your body. But you don't move anything when you just think. So, to a scientist, thinking is not work.

Every time you move something, you either push or pull it. If you slide a pencil away from you, you push it. If you pick up a pencil, you pull it. Both times you are doing work to the pencil.

Both pushing and pulling are kinds of **force**. Force is what is needed to do work. So here is an even better science definition for work: Work is using force to move something from one place to another.

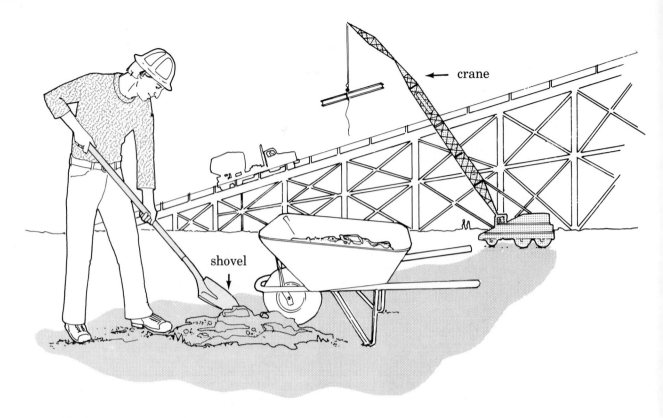

crane

shovel

Machines

Lift this book from your desk. You're doing work because you're using force to move the book.

It's easy work to lift a book. You don't use much force. Suppose you lift a small box of books. Now the work is not so easy. You must use a lot more force.

What if the box of books is huge? It's so heavy that it won't budge when you try to lift it. How can you move that box?

One way is to use a **machine**. A machine is anything that can help you do work. Machines are helpful because they let you do hard work with less force. Think of some machines that help you do work. What are they?

You may have thought of machines such as a car or washing machine. These machines use gasoline or electricity as sources of energy for work.

But a machine might have no source of energy other than a person's muscles. A machine can be a **tool** such as a knife, a hammer, or a potato masher. Tools help you do work with much less force than you would need without them.

A hammer helps you drive a nail into wood. Could you do that with just your hands? A potato masher helps you use enough force to mash a potato into mush.

The picture above shows workers building a freeway ramp. The workers use machines to help them work. One is a large machine called a *crane*. It runs on fuel. What work does a crane do?

The workers are also using some tools, such as shovels. What work does a shovel do? How does it make this work easier?

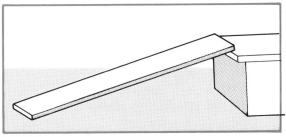

Inclined plane

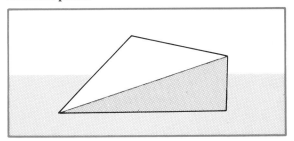

Wedge

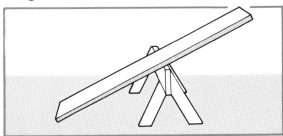

Lever

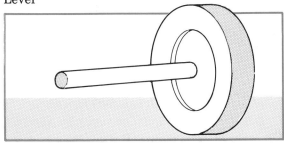

Wheel and axle

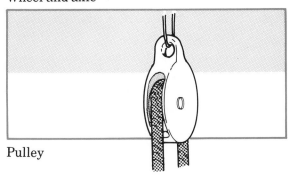

Pulley

Simple Machines

Machines come in different shapes and sizes. They do different kinds of work. Some are small and some are large. But all machines are made up of certain basic parts. These parts are called **simple machines**.

The pictures on this page show some of the most common simple machines. Each of these simple machines can be combined with other simple machines. Together, they can make larger machines.

What are the names of these simple machines?

Think of some kinds of work each simple machine can do. What work can an inclined plane help you do? What work can a lever help you do?

Look at the picture on the opposite page. What machines use a wheel and axle? What machines use a lever? pulley?

Review

Use what you learned in this unit. Finish the sentences. The words you'll need are listed below.

force	push	less
machines	more	tools
simple machines	work	

1. You either _____ or pull something when you move it.
2. You use _____ when you push or pull something.
3. When you do easy work, you use _____ force.
4. When you do hard work, you use _____ force.
5. Things that help you do work are called _____.
6. All machines are made of _____ _____.
7. You do _____ whenever you move something.
8. Machines such as a hammer and shovel are also called _____.

Check These Out

1. Make a Science Notebook for Machines. Put your list of glossary words and their meanings in the notebook. Also keep a record of the experiments you do. You can put anything else you learn about machines in your Science Notebook too.
2. Look through magazines and newspapers. Cut out pictures of machines. Make a poster with them.
3. What jobs do you do around the house? Write down those jobs. Write the names of machines that can help you do your work.
4. What would our world be like if we didn't have machines? Think about it. Then make up a story. Tell or write your story.
5. Choose a machine and find out more about it. Find out who invented the machine and when it was invented. What did it look like then? What does it look like today? How does that machine help people do work? Give a report about that machine.
6. As you work through this section, you may want to find out more about machines. You can find out more by looking in an encyclopedia or by getting books from a library. You can also talk to an expert, such as an auto mechanic, a shop teacher, an engineer, or a physical scientist.

   Here are some things you may want to find out:
   - What is mass? What is weight?
   - What is gravity? How does gravity affect how much force you use to move an object?
   - What was the Industrial Revolution? How did it change our history?

Unit 2

The Inclined Plane

Which is easier: *carrying* a heavy box or *pushing* it across the floor? Pushing the box would probably be easier. Now suppose you had to place the box on a high shelf. Lifting that box would be even harder than pushing it.

If only you could just push something from a floor to a high place. Work would be so much easier. One simple machine can help you do just that.

- What is that simple machine?
- How does it help make your work easier?

You'll learn the answers in this unit.

Before You Start

You'll be using the science words below. Find out what they mean. Look them up in the Glossary. On a separate piece of paper, write what the words mean.

1. **ramp**
2. **slope**
3. **steep**

Ramps

Picture this:

You're a member of a band. And you're loading equipment into a van. The loudspeakers are too heavy to lift. Is there a way to push the loudspeakers into the van?

Yes! You can use a *ramp*. The ramp is an **inclined plane**—a simple machine.

To make the ramp, you get a long board. You lean one end of it against the van. The board is now a ramp! You put the loudspeakers on the ramp and push them up and into the van.

Remember: A machine can help you do work. How do you think inclined planes make work easy? (Hint: Look at the picture.)

An inclined plane connects a low place to a high place. So it lets you slowly push or pull something up to the high place. You use less force moving it that way than if you carry it.

Ramps or inclined planes are easy to make. All you need is a board. Think of a time when you had to lift a heavy thing. Could a ramp have helped you then? What thing were you trying to lift?

Concrete or wood ramps are often built into streets and buildings. For example, ramps are built for disabled people. How do you think those ramps make work easy?

How Much Force?

A machine makes your work easier. Find out how an inclined plane makes work easy. You'll need these things:

- One board, 1 foot long
- Two thick books
- One heavy coffee cup
- One thin rubber band

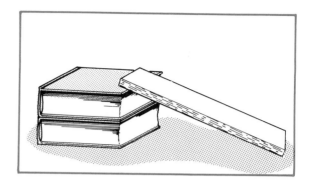

1 First make a ramp. (Remember: A ramp is an inclined plane.) Stack the books. Then lean one end of the board on the books.

2 Loop the rubber band around the cup handle, the way the picture shows. The rubber band will show how much force you use to move the cup. The more the rubber band stretches, the more force you use.

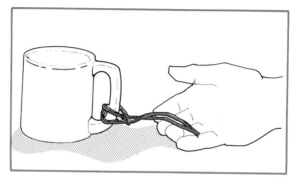

3 Put the cup next to the books. Hold the rubber band and lift the cup. Lift it to the top of the ramp.

Look at the rubber band. Does it stretch a lot or a little?

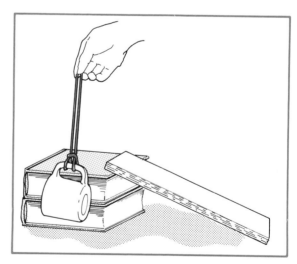

4 Now put the cup at the bottom of the ramp. Pull the rubber band and slide the cup up the ramp. Does the rubber band stretch more or less than before?

The rubber band should stretch less.

When do you use less force: when you pick up the cup or when you slide it up the ramp?

Experiment 1

What kind of inclined plane makes work easier?

Some inclined planes are steep. Some are not. Which kind makes work easier? Do this experiment and find out.

Materials (What you need)

Twelve thick books (all the same size and thickness)

One coffee cup with a rubber band tied to its handle

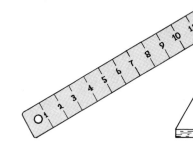

One ruler that has a hole in it

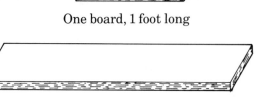

One board, 1 foot long

One board, 3 feet long

Procedure (What you do)

1. Make two ramps: Stack six books and lean the short board against them. Stack the other books and lean the long board against them.

2. Put the cup on the bottom of the short ramp. Put the end of the rubber band through the hole in the ruler.

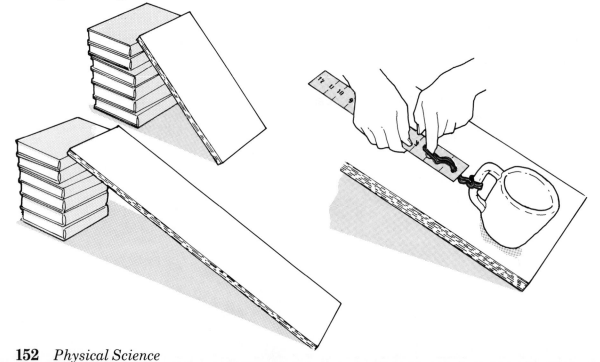

3. Pull the rubber band and slide the cup up the ramp. Look at the ruler as you pull. How many inches does the rubber band stretch?

4. Put the cup on the long ramp. Pull it up as before. How many inches does the rubber band stretch?

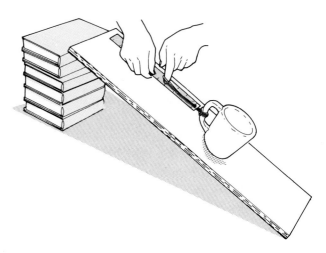

Observations (What you see)

1. Which makes a steeper slope: the short board or the long board?
2. Where does the rubber band stretch less: on the short ramp or on the long ramp?

Conclusions (What you learn)

1. When do you use more force?
   a. Moving things up a slope that's very steep
   b. Moving things up a slope that's not so steep
2. When is work easier?
   a. Moving things up a very steep slope
   b. Moving things up a less steep slope
3. Suppose you're making a ramp. Which makes your work easier?
   a A short ramp
   b. A long ramp

Machine Choice

You learned that inclined planes can help make your work easier. Read these two problems. Decide which inclined plane is the better choice.

1 Picture this:

Your job is to build a ramp to the door of a building. The ramp is for people in wheelchairs. They need the ramp to get into the building.

Picture A and picture B show two ramps you could make. Which ramp will be easier for people in wheelchairs to use?

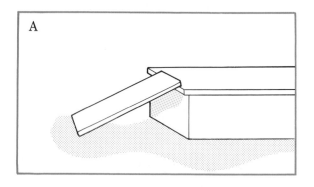

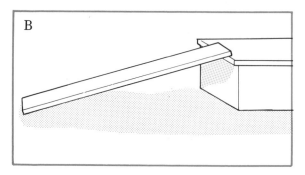

2 You and a friend want to go to the building on the hill. You decide to walk up the cement path. Your friend walks up the stairs.

Who will get to the building first? Why do you say that?

Who will do the most work getting to the building? Why do you say that?

Who made the better choice? Explain your answer.

Review

Show what you learned in this unit. Find the right words to finish the sentences.

1. An inclined plane can make your work
   a. easier.
   b. harder.
   c. difficult.
2. An inclined plane connects
   a. a low place to a high place.
   b. two places that are the same height.
   c. two things that are moving.
3. An inclined plane helps you
   a. cut things.
   b. split things.
   c. lift things.
4. When an inclined plane is steep, you use
   a. less force.
   b. more force.
   c. no force.

5. Look at the picture. It shows two inclined planes. One is circled. Find the other inclined plane.

Check These Out

1. Find pictures of inclined planes. Look through magazines and newspapers. Cut out the pictures. Then make a poster with those pictures.
2. How do you think the first inclined plane was invented? Write a story about what you think happened.
3. Many roads go up steep hills and mountains. Invite someone who builds roads to your class. Ask that person to talk about the problems there are in building those roads.
4. The pyramids of Egypt were built more than 4,500 years ago. Find out how people used inclined planes to make them. Draw a picture of a pyramid. Tell how it was built.
5. Here are more things you may want to find out:
   - Why is the screw an inclined plane? What are some examples?
   - What amusement rides have inclined planes?
   - Why is an escalator an inclined plane? How does it work?

The Wedge

You cut food with a knife. You open cans with a can opener. The knife and the can opener are both machines. They help you do some kind of work. You learned that machines can be made up of simple machines. Both the knife and the can opener have parts that are **wedges**. The wedge is a simple machine.

- What is a wedge?
- How does it make your work easier?
- What jobs can you use a wedge for?

You'll learn the answers in this unit.

Before You Start

You'll be using the science words below. Find out what they mean. Look them up in the Glossary. On a separate piece of paper, write what the words mean.

1. **concentrate**
2. **split**

Pushing Things Apart

Suppose you're chopping wood for a fire. You chop long pieces of wood into short, thin pieces. What tool do you use?

Right! You use an ax. The ax has a part that can split things. That part is a simple machine—the *wedge*.

Look at the picture of a wedge. How would you describe its shape?

The wedge is a triangle. One end is wide. The opposite end is pointed. Because the wedge has that shape, it can move through things. It can push them apart.

For instance, you use a knife to cut an apple in half. The sharp part of the knife is a wedge. It cuts into the apple. Then it pushes the two sides of the apple apart.

Axes and scissors also have wedges. What other tools have wedges?

Some wedges can help you lift things and hold them in place. A doorstop is a kind of wedge. When you put it between a door and the floor, this happens: It pushes the door up a little. And it holds the door in one place.

Where else could you put a wedge to lift something and hold it in place?

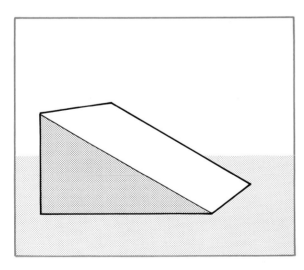

A wedge

Wedges Make You Stronger

It is very hard to split things apart. A wedge can help you do that. That's because a wedge *concentrates* the force you use. It puts that force in a very small place. It makes you stronger than you are.

See how a wedge concentrates your force. You'll need these things:

- Two soup cans
- Masking tape
- One book
- One wedge, such as a doorstop

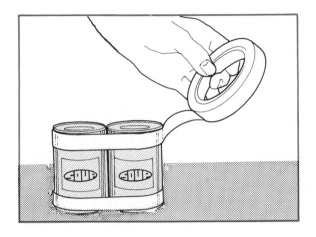

1 Tape the two cans together. Tape them all around the top and bottom.

2 Hold one can in each hand. Pull on the cans. What happens?

Right! Nothing happens. No matter how hard you pull on the cans, you won't split them apart.

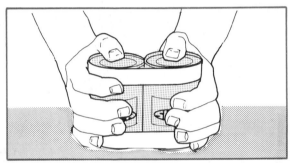

3 Now lean the cans on the book, the way the picture shows.

4 Put the pointed end of the wedge between the two cans. Push the wedge completely between the cans. What happens?

Right! The wedge concentrates your force—and you split the cans apart!

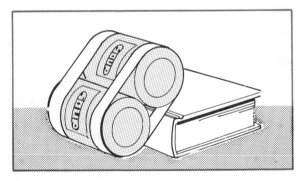

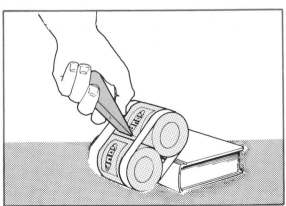

Different Shapes

Wedges have different shapes. We use wedges to do different kinds of work. One kind of wedge makes holes. And another kind cuts things. What does a third kind of wedge do?

That's right! It splits things apart.

Many different machines have wedges. Very large machines called *backhoes* have wedges. Backhoes are used on jobs such as building freeways. The wedges on a backhoe help the workers dig into the earth and split rocks.

Smaller machines, such as a lawn mower, also have wedges. What do the wedges on a lawn mower do?

Right! They cut the grass.

Some small tools, such as pins and needles, also have wedges. Pins and needles don't cut or split things apart. What do they do?

Right! They make holes in things.

Here are some other machines that have wedges. What kind of work do they help you do? Do they cut, split apart, or make holes?

1. scissors
2. knife
3. nail
4. chisel
5. ax

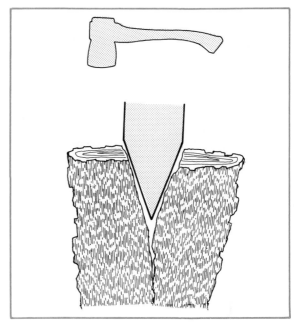

Splits apart

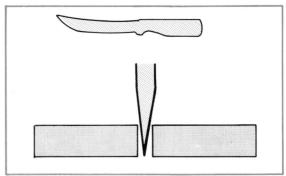

Cuts

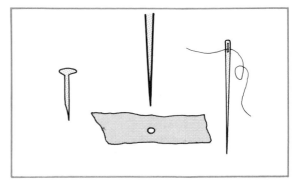

Makes holes

Machine Choice

The pictures on the right show some tools that have wedges. Different workers use them. Look at the pictures and answer the questions below. Then check your answers. (The right answers are upside down.)

1. You're a cook's helper in a restaurant. You cut many vegetables. Which tool do you use?
2. You work at a gas station. You are cutting a hole in a can of motor oil. What tool do you use?
3. You work in a grocery store. You open a wooden box that's nailed tightly shut. You use two tools to open it. Which tools do you use?
4. You work in a clothing store. You are sewing the hem of a skirt. Which tool do you use?

Answers

4. needle
3. hammer and chisel
2. can opener
1. knife

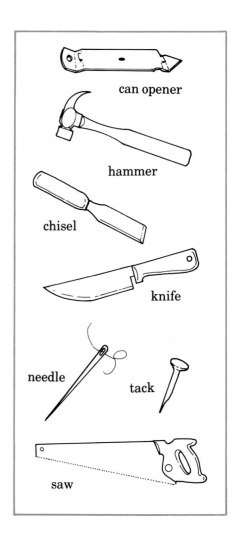

can opener

hammer

chisel

knife

needle

tack

saw

Review

Show what you learned in this unit. Find the right words to finish the sentences. There may be more than one correct answer for a sentence.

1. A wedge is a simple machine that
   a. is round.
   b. is like a triangle.
   c. is square.
2. Wedges can be used to
   a. cut things.
   b. split things.
   c. make holes.
3. When you use a wedge, you
   a. use less force.
   b. concentrate your force.
   c. use no force.
4. Wedges are part of
   a. large and small machines.
   b. tools.
   c. steps.

5. Look at the picture. It shows two wedges. One wedge is circled. Find the other wedge.

Check These Out

1. Many sharp wedges, such as those on knives and scissors, can hurt you. What are some safety rules for using sharp wedges? Write them down on a separate piece of paper.
2. Find pictures of wedges. Look through magazines and newspapers. Cut the pictures out and paste them on poster paper.
3. Why is a rowboat shaped like a wedge? Find a picture of a rowboat and bring it to class. Start a discussion with your classmates about the boat's shape.
4. Some wedges, such as needles, are used to make holes. Find out what other wedges do that. Make a list of those wedges.
5. Here are more things you may want to find out:
   - What did the first tools that people made look like?
   - What are some machines that have wedges?
   - Who was Cyrus McCormick? What big machine did he invent? How are wedges important parts of that machine?

Unit 4

The Lever

Did you ever take a tire off a car?

If you did, you probably used three different tools. You used one tool to take off the hubcap. You used another tool to unscrew the nuts that hold the wheel onto the car. And you used one more tool to help you lift the car.

Those three tools are all examples of one kind of simple machine: the **lever**. Levers let you do very hard work that you are not strong enough to do.

- What is a lever like?
- How does a lever work?
- When should you use levers?

You'll learn the answers in this unit.

Before You Start

You'll be using the science words below. Find out what they mean. Look them up in the Glossary. On a separate piece of paper, write what the words mean.

1. **fulcrum**
2. **increase**
3. **object**

Getting Stronger

How strong are you? Can you take off a wheel nut with just your fingers? Carry 300 pounds of rocks in your arms? Open a can of paint with just your fingernails?

Most of us can do just so much hard work. That's because our muscles can push or pull with only so much force. And it isn't enough force to move something that's tightly stuck together, such as a nut and bolt, or the lid on a can of paint.

To do those kinds of hard work, we use levers. Levers can make us stronger than we are.

There are many different kinds of levers. Some of them look very different from each other. But all levers use a *fulcrum*. A lever cannot work without a fulcrum. Look at the lever in the top picture. What part is the fulcrum?

Levers work this way: You rest the lever against a fulcrum. You push one end of the lever against the fulcrum. That makes the other end of the lever push whatever is on it or next to it. That end moves an object with greater force than the force you use to push the lever.

Look at the pictures of tools. They all are machines that are also levers. See if you can find their fulcrums—the parts that you push the levers against.

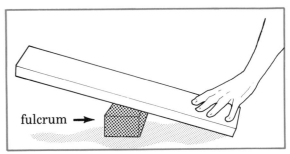

lever

The wheelbarrow is used to carry heavy loads.

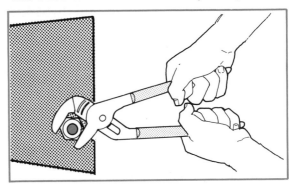

Pliers can be used to take off nuts.

A crowbar can move something that is stuck.

Experiment 2

Which makes you stronger: a short or a long lever?

Think about this: You can lift a person with only your thumbs! All you need is a board and a book. What do you do with those things?

Right! You make a lever.

When you use a lever, you can sometimes lift objects with very little force. That's because a lever *increases* your force.

Can a short lever increase force more than a long lever? Get a partner and find out.

Materials

One thick book, such as a desk dictionary

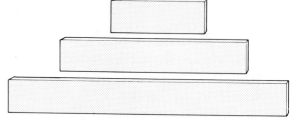

Three boards: 2 feet, 4 feet, and 6 feet long

Procedure

1. Lay the shortest (2-foot) board on the book. The book should be under the center of the board.

2. Your partner stands on one end. With your hands, push down on the other end. Can you lift your partner?

3. Lay the next longest (4-foot) board on the book. The end should be 1 foot from the book. Have your partner stand on the end that's near the book. Can you lift your partner?

4. Now do the same with the longest (6-foot) board. The end should be 1 foot from the book. Can you lift your partner?

Observations

1. Which lever first lifts your partner?
2. Which lever is the easiest to use? (Hint: Think of the one you push with the least force.)
3. Which lever do you push with the most force?

Conclusions

1. Which lever lets you use less force: a short or long one?
2. Which lever increases your force more: a short or long one?

Where Do You Put the Fulcrum?

All levers have a fulcrum. You can put that fulcrum anywhere on the lever. Find out what happens when you change the place you put the fulcrum. You'll need a partner. You'll also need these things:

- One thick book
- One board 6 feet long

1 Lay the board on the book. The book should be 1 foot from the end of the board. The book is the fulcrum.

2 Your partner stands on the end that's *farthest* from the fulcrum. With your hands, push down on the other end. What happens?

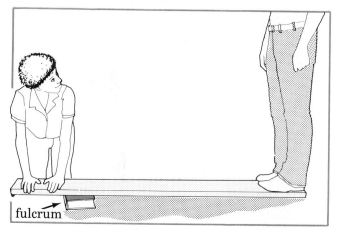

3 Now place the book beneath the center of the board. Your partner stands on the same end again. Push down on your end. What happens?

4 Place the book so it's 1 foot from the end that your partner stands on. Push down on your end. What happens?

Where should you put a fulcrum when you have an object that's hard to move: near or far from that object?

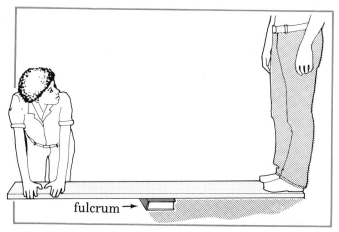

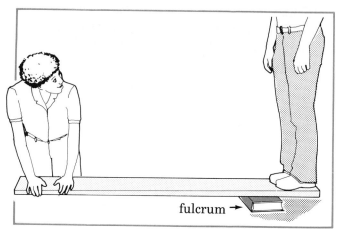

Machine Choice

Picture this:

You and your friends have been camping. Now you're driving home.

Suddenly you stop. A huge rock has fallen in the middle of the road! You can't get around it. How can you move that rock out of the way?

You have a car jack and a crowbar in the car trunk. And on the ground you see a long, heavy branch that fell off a tree.

You can use the car jack, the crowbar, or the branch as a lever. Which one do you pick? Why?

You find a smaller rock to use as your fulcrum. Where would you put it? In the picture at the right, find where you would put the fulcrum.

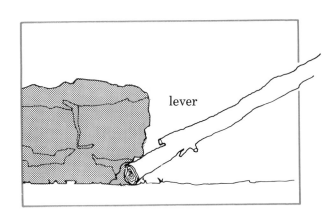

lever

Review

Show what you learned in this unit. Find the right words to finish the sentences. There may be more than one correct answer for a sentence.

1. Levers help you
    a. lift heavy objects.
    b. move objects that are stuck.
    c. split objects.
2. A fulcrum is
    a. another name for a lever.
    b. the part a lever moves on.
    c. the object you move.
3. When a fulcrum is close to an object, you
    a. use less force to lift it.
    b. use more force to lift it.
    c. can't lift it.

4. Look at the picture. It shows three levers. One lever is circled. Find the other two levers.

Check These Out

1. One kind of can opener is a wedge and a lever. That kind of lever is sometimes called a *church key*. Bring one to class. Explain why it is a wedge and a lever. Find the fulcrum. Explain how the can opener works.
2. Many large machines, such as cranes, have levers. Find pictures of large machines that have levers. Look in magazines and newspapers. Put the pictures on a bulletin board. Ask others to find the levers.
3. What levers do you have at home? Make a list.
4. Bring examples of levers to class. Give a report about how each one works.
5. Here are more things you may want to find out:
    *   What does *mechanical advantage* mean?
    *   How is your arm like a lever?
    *   What are compound levers? What are some examples?

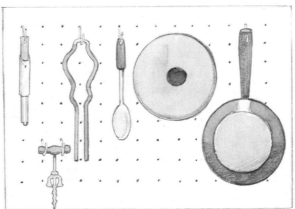

Unit 5

The Wheel and Axle

Riding a bus is easier work than walking.

Using an eggbeater to mix batter is easier than beating the batter with a fork.

The bus and the eggbeater both make work easier. They are machines that have many parts. One of those parts is the simple machine you'll be learning about. That machine is the **wheel and axle**.

- What are the wheel and axle like?
- How do they make work easier?
- How do they make parts in machines move?

You'll learn the answers in this unit.

Before You Start

You'll be using the science words below. Find out what they mean. Look them up in the Glossary. On a separate piece of paper, write what the words mean.

1. **control**

2. **gear**

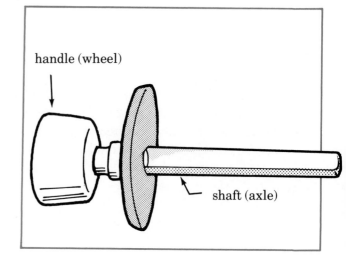

A doorknob

handle (wheel)

shaft (axle)

Moving in Circles

Look at the picture of the tightly closed door. Could you open a door like that just by pushing it?

You wouldn't be able to open that door, no matter how hard you push it. That's because a metal part inside the door—the *latch*—sticks out of the door and into the door frame. To open the door, you must move the latch. You do that by using a simple machine called the *wheel and axle*. What door part do you think is a wheel and axle?

Right! The *doorknob* is a wheel and axle.

The wheel and axle is made up of two kinds of parts. One part turns in a circle like a wheel, so it is called the *wheel*. The other part is attached to the wheel and moves in a smaller circle in the center of the wheel. That part is called the *axle*.

When you turn the wheel, you cause the axle to turn also. What do you think happens when you turn the axle?

Right! The wheel also turns.

Look at the picture of the doorknob. It shows the two kinds of parts that make up that simple machine. The handle of the doorknob is the wheel. The shaft is the axle. The shaft is placed next to the latch. When you turn the handle, you turn the shaft—and that moves the latch so you can open the door.

Here are some other examples of wheels and axles that work like the doorknob: radio knobs, water faucet handles, windmills, and the controls on appliances, such as gas stoves. Each has a part that turns like a wheel and a part connected to the wheel that is an axle. When the wheel or the axle turns, it causes the other part to turn also.

Sometimes you'll see wheels and axles on things such as roller skates and shopping carts. Those wheels are not connected to the axle. The wheels turn around the axle, but the axle does *not* turn. Do you think those wheels and axles are simple machines? Why?

Increases Our Force

In Unit 4 of this section, you learned that the lever increases your force. It helps you move objects that are difficult to move. A wheel and axle also increases your force so that you can move things.

In fact, a wheel and axle is like a lever that moves in a circle. You push against a part that moves in a large circle. That causes a second part to move also in a small circle inside the large circle. That second part moves the object that is connected to it.

A wheel and axle helps you do things you couldn't do with just your own force. It helps you haul and hoist heavy objects, move objects that are tightly stuck, and push objects.

You've probably used a wheel and axle and not realized that you have. That's because many wheels and axles do not look like a wheel. A wheel and axle does not need to *look* like one. It just needs to move like a wheel.

For instance, a wrench is a wheel and axle. You use it to tighten or to take off nuts and bolts.

Look at the picture of a wrench. Notice that it is made up of a handle and a part that fits snugly on a bolt or nut.

Let's say you're taking a nut off a bicycle wheel. In order to do that, you fit the wrench on the nut. Then you turn the handle in a circle until the nut comes off.

Which part of the wrench moves in a large circle?

Right! The handle moves in a large circle. That part is the wheel. The part that fits on the nut moves in a small circle. That part is the axle.

Suppose you didn't have a wrench to take the nut off. Could you do it with your bare hands? Why or why not?

You can see how a wheel and axle lets you do work you couldn't do by yourself. Get a jar that's so tightly shut you can't open it with your hands alone. Get a jar opener and use it to open the cover.

How does the jar opener work like a wheel and axle?

Some wheels and axles

wrenches   jar lid openers

screwdrivers

handles on other machines

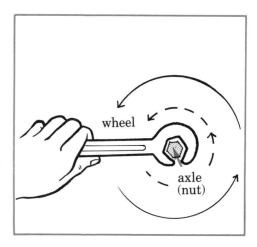

wheel

axle (nut)

You can use a wrench to take off a nut.

Another Kind of Wheel and Axle

You've probably seen watches with hour and minute hands. Have you ever looked inside one? If you did, you would see several tiny gears. Those gears are a kind of wheel and axle.

Many machines—large and small—have gears. Gears move parts in a machine. For example, gears move the hands in a watch.

Here's how gears work: A gear turns another gear. That gear is joined to a part in a machine. As the gears move, that part moves too.

Gears can turn each other because they have *teeth*. Look at the picture of the gears at the top of the page.

Now make a model of two gears, and see how they work. You'll need these things:

- Two bottle caps (the kind that you take off with a bottle opener)
- One block of soft wood (such as pine)
- Two push pins
- One hammer

1 Put a bottle cap face down on the wood. Put a push pin in the middle of the cap, and hammer it into the wood.

2 Place the other bottle cap so it tightly touches the first one. Put the other push pin in the middle of the cap. Hammer it into the wood.

3 Turn one bottle cap. What happens?

Right! The other bottle cap turns too. Why does it turn?

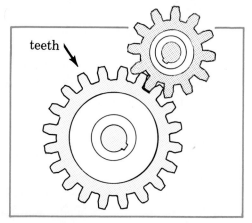
teeth

A gear can turn another gear.

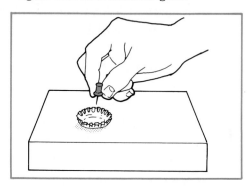

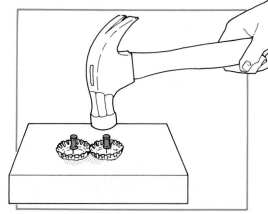

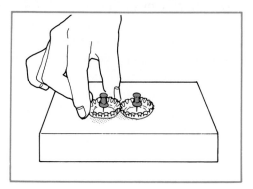

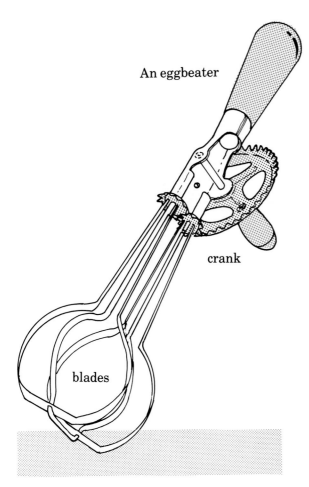

An eggbeater

crank

blades

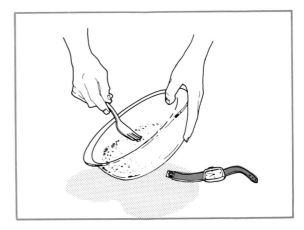

A Close Look at Gears

Gears move parts of machines. Gears also control how fast or slow parts move. You can see that with an eggbeater.

Look at the eggbeater above. It has a large gear and small gears.

Now get an eggbeater like the one in the picture. Turn the large gear with the crank. What happens to the small gears next to it?

Right! The large gear turns the small gears.

A blade is attached to each small gear. The gears move those blades quickly or slowly. Try turning the crank so the blades move slowly. Then turn it faster, and even faster.

Now see how the eggbeater can make work faster and easier for you. Get half a cup of whipping cream (or two egg whites), a fork, and a watch. Put the whipping cream in a bowl. Using the fork, beat the cream until it is fluffy and too thick to be poured. Look at your watch when you start and when you finish. How long does it take?

Get another half-cup of whipping cream, and put it in a bowl. Now use the eggbeater to beat the cream. Time yourself. How long does it take to make the cream thick and fluffy?

Machine Choice

When you sharpen your pencil with a hand-turned pencil sharpener, you are using a wheel and axle. Look at a pencil sharpener like the one in the picture. Take off its case. Then look for these parts: pencil holder, cutting rollers, gears, and handle.

1 Turn the handle. What happens to the other parts?

2 Which part of the pencil sharpener is the wheel? (Hint: It looks something like the spoke of a wheel.)
Why do you say that part is the wheel?

3 Which part is the axle?
Why do you say that part is the axle?

4 What causes the cutting rollers to move in a circle?

5 What causes the cutting rollers to spin?

6 Now put a pencil into the pencil holder. Turn the handle a few times. Explain how the parts work together to sharpen the pencil.

A hand-turned pencil sharpener

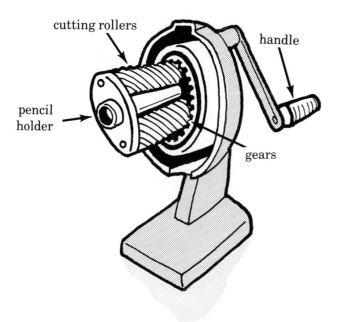

cutting rollers

handle

pencil holder

gears

Review

Use what you learned in this unit to answer the questions. The page number after each question shows where you can find the answer.

1. What kind of work can a wheel and axle help you do? (page 171)
2. Why does a wheel and axle make it easy to move things? (page 171)
3. What makes a gear turn another gear? (page 172)
4. Why are gears used? (page 173)
5. The picture shows three machines that have wheels and axles. One machine is circled. Find the other two machines.

Check These Out

1. Invite an auto mechanic to class. Ask him or her to speak about the different kinds of wheels and axles on a car.
2. Find pictures of machines that have wheels and axles. Look in magazines and newspapers. Cut the pictures out and make a poster.
3. Get the model of gears that you made on page 172. Add more gears to it. Does each gear turn? Do they all turn the same way?

4. Here are more things you may want to find out:
   - What is torque?
   - What is a windlass? How does it help you do work?
   - What are some wheels and axles that you use for play and work?

Unit 6

The Pulley

Did you ever ride on a Ferris wheel? A simple machine makes the Ferris wheel turn. That same simple machine is used to raise and lower the huge tents at a circus. What's the machine? A **pulley**.

- What is a pulley like?
- How does it make work easier?
- How does it make other machines work?

You'll learn the answers in this unit.

Before You Start

You'll be using the science words below. Find out what they mean. Look them up in the Glossary. On a separate piece of paper, write what the words mean.

1. **groove**
2. **pulley system**

Belts, Ropes, and Pulleys

Picture this:

You're using a vacuum cleaner. But the roller brush isn't picking up dirt. So you turn the vacuum cleaner off and open it up. *Aha!* The belt that connects the roller brush to the motor is broken!

You get a new belt. Now where do you put it? Look at the diagram. Find the place where the belt should go.

There's a certain kind of wheel on the motor. That wheel has a *groove*. The roller brush also has a groove. The wheel with the groove is a pulley. The brush with the groove is also a pulley.

A pulley is a wheel with a groove around it. You put a belt or rope around the wheel. The groove keeps it on the wheel. You use pulleys to move things like the brush in the vacuum cleaner.

You can also use a pulley to lift heavy things. Here's how: You tie one end of the rope to the thing you want to move. You pull the other end of the rope and that thing goes up.

When two or more pulleys are used, they make up a *pulley system*. Many different workers use rope and pulley systems. Those systems help them lift heavy objects. For example, painters use pulley systems to lift themselves up tall buildings.

The two pulleys and the belt in the vacuum cleaner also make up a pulley system. Machines with moving parts often have pulley systems like that. Motors are attached to pulley systems inside those machines. What's one machine that has a motor attached to a pulley system?

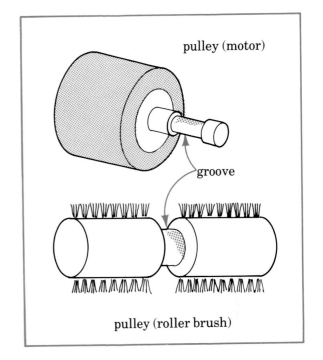

pulley (motor)

groove

pulley (roller brush)

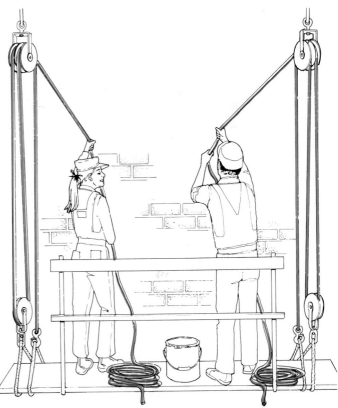

Painters use pulley systems to lift themselves up sides of buildings.

Why Use More than One Pulley?

You learned about pulley systems. A pulley system is made up of two or more pulleys. Why do we use more than one pulley? Find out.

You'll need these things:
- Two boards: 3 feet and 4 feet long
- Two strong tables, placed about 2 feet apart
- Fifteen feet of thin rope
- Scissors
- Strong string
- Three curtain pulleys
- Some thick books (such as desk dictionaries)

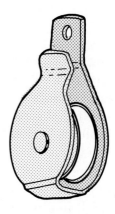

curtain pulley

---

1 Lay the long board across the tables. Lay the short board on the floor, under the long board.

2 Cut some rope and tie it around the long board. With string, tie two pulleys to the rope.

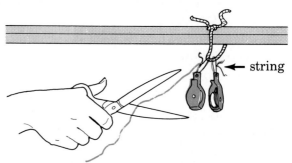

← string

Cut more rope and tie it around the middle of the short board. With string, tie one pulley to that rope.

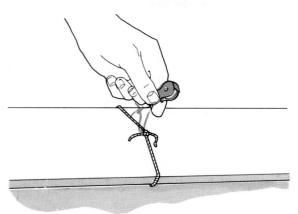

3 Tie one end of the rope to the short board.

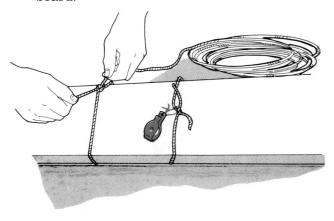

4 Loop the other end around one of the top pulleys.

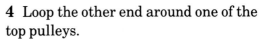
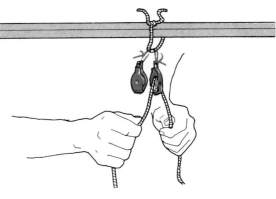

5 Put the books on the short board. Pull down on the rope. Is the board easy to lift?

6 Loop the end of the rope around the bottom pulley. Then loop the rope around the top empty pulley. Pull down on the rope. What happens?

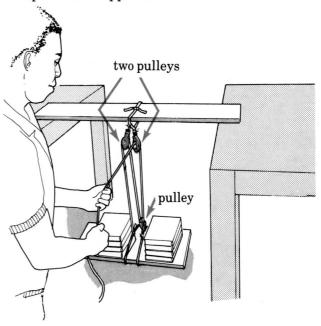

two pulleys

pulley

What Happens?

1. What happens when you use one pulley?
2. What happens when you use three pulleys?

Moving Parts

Pulley systems are part of many machines. They are used to move parts of machines.

Make a model of a pulley system, and see how it works. You'll need these materials:

- Two lids from mayonnaise jars
- One block of soft wood (such as pine)
- Two push pins
- One hammer
- One 3-inch rubber band
- One mayonnaise jar

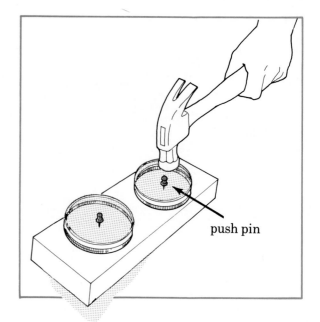

push pin

1 Place the lids on the block of wood. They should be 1 inch apart. Hammer a push pin into the middle of each lid.

2 Loop the rubber band around the two lids. Turn one lid. What happens to the other lid?

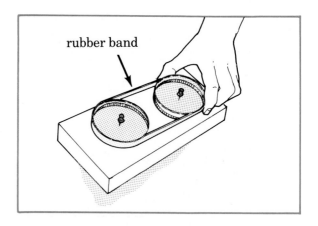

rubber band

3 Screw the jar to its lid. If you turn the other lid, what do you think will happen to the jar? Try it. What happens?

That's right! The jar turns. Every time that lid turns, the jar also turns.

A pulley system in a machine works in that way. A moving part is attached to one pulley. When the other pulley turns, the part moves.

Some machines have parts that move around and around in pulley systems. An escalator is an example. What other machine is like that?

Machine Choice

Imagine this:

You and a friend are garden workers. You're taking out a small tree.

First you cut the tree down. You use an ax to do that. Then you and your friend start digging out the tree stump. But you need something to help you take it out.

Look at the pictures of different machines. Which machine (or machines) would you use?

How would you use the machine or machines?

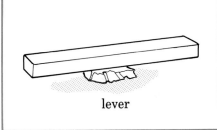

lever

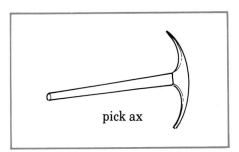

pick ax

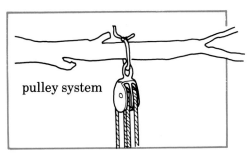

pulley system

Review

Use what you learned in this unit to answer the questions. The page number after each question shows where you can find the answer.

1. What is a pulley? (page 177)
2. What is a pulley system? (page 177)
3. What happens when you add more pulleys to a pulley system? (page 179)
4. What do pulley systems do in machines? (page 180)
5. Look at the picture. It shows two pulley systems. One pulley system is circled. Find the other one.

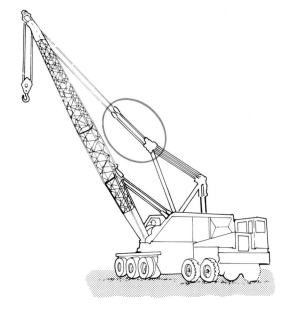

Check These Out

1. A cannery is a factory that puts food into cans. In some canneries, workers stand at *conveyor belts* and put food into cans. Find out what a conveyor belt is and how pulley systems make it work. Give a report about conveyor belts.
2. Visit a factory that uses conveyor belts.
3. Film projectors use pulleys. Look at a projector. Find out where the pulleys are and what they do. Make a drawing that shows how the film projector works.
4. Look through magazines for pictures of machines that have pulley systems. Cut the pictures out and make a poster with them.
5. Here are more things you may want to find out:
   - What is a fixed pulley? A moveable pulley?
   - Where do you find pulleys on ships?
   - How do escalators work?
   - How do elevators work?

Unit 7

Using Machines

You learned about five simple machines. What are they?

Simple machines can be used by themselves. They can also be put together to make other machines.

The pictures show two machines that are made up of two or more simple machines. Find the simple machines that make up each machine. Then, on a separate piece of paper, write the names of the simple machines that you found. Make a list for the lawn mower and for the backhoe.

lawn mower

backhoe

Working in a Warehouse

Suppose you work in a warehouse that stores and ships books. You do different kinds of work each day. On these pages, read about some of the work you do. Then decide what machines you would use.

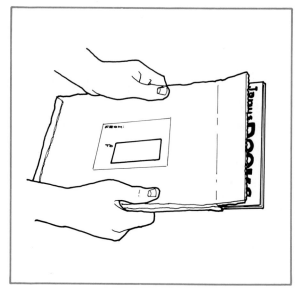

Opening Boxes, Packing Books

1. You are opening boxes. They are taped with heavy paper tape. The pictures on the right show what machines you could use to open them. Which one would you use?

   What simple machines are part of it?
2. You pack books in bags. You must seal the bags tightly shut. Look at the pictures on the right. What machine would you use to seal the bags?

   What simple machines are part of it?

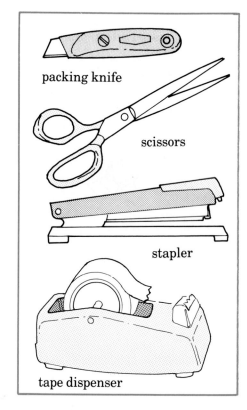

packing knife

scissors

stapler

tape dispenser

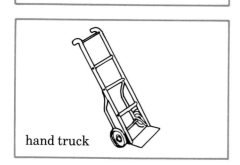

forklift

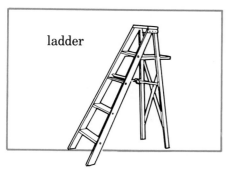

hand truck

ladder

Putting Books Away

1. Your next job is to put boxes of books on shelves.
    The pictures on the left show the machines you
    could use. (If you're not sure what those machines
    do, look them up in the Glossary.) Suppose you put
    boxes on a shelf that's one foot above the ground.
    What machine would you use?

    What simple machines are part of it?
2. Sometimes you put very big, heavy boxes on
    shelves that are 15 feet high. What machine would
    you use to help you?

    What simple machines are part of it?

Show What You Learned

What's the Answer?

Simple machines help you do many different kinds of work. Pick the right words from the list below to answer the questions.

inclined plane   wedge   lever
wheel and axle   pulley

1. What helps you push or pull things up from the ground to a higher place?
2. What helps you move things that are tightly stuck? Or lift heavy things?
3. What helps you lift things up to a very high place?
4. What helps you cut, split, and make holes?
5. What turns in circles and helps you move things?
6. What simple machines increase force and make you stronger? (Name three.)

What's the Word?

Give the correct word or words for each meaning.

1. A push or pull
   F _____
2. When you move something from one place to another
   W _____
3. To put all a force in one very small place
   C _____
4. All machines are made of at least one of these.
   S _____ M _____
5. One kind of inclined plane
   R _____
6. To make more
   I _____

Congratulations!

You've learned a lot about machines. You've learned

- What all machines are made of
- What the five simple machines are
- How each simple machine can help you do work
- And many other important facts about machines

Glossary
Energy

ap pli ance   Something that uses energy to give us heat, light, or motion

clean en er gy   Energy that does not make the air dirty

coal   A rock that is burned for energy

con serve   To use something wisely

crude oil   A thick black liquid that is burned for energy; petroleum

en gine   A machine that burns fuel and gives us energy

ex haust   Smoke that comes out of an engine

fos sil   The remains of a plant or animal that lived long ago

fos sil fuels   Materials we burn that are found under the ground

fu el   Material we burn for energy

gen er a tor   A machine that makes electricity

heat col lec tor   A thing that gets hot when the sun shines on it; a part of a solar water heater

in su la tor   A material that keeps heat from leaving quickly

mag net   A material that pulls on iron and steel

mag ne tism   The force that magnets have

me ter   A machine that measures how much electricity is used

mine   A place where rock such as coal is taken out of the ground

mo tion   What happens when something moves

nat u ral gas   A gas found underground that is burned for energy

nu cle ar fuel   A material that can give off heat without burning

pol lute   To make air, water, or land dirty

po ten tial en er gy   The energy that is held inside of things

pow er plant   A place that makes electricity for a community

pro duce   To make something

prod ucts   Things that are made

ra di o ac tive   Giving off dangerous rays or tiny bits of materials

re fin er y   A factory that makes gasoline and other things from crude oil

re new a ble   A source of energy that will not run out

ro tor wheel   A wheel that's turned in order to make electricity

sil i con   A material made from sand. *Silicon is used to make solar cells.*

so lar cell   A thing that changes the sun's energy into electricity

stored   Kept inside of something

u ra ni um -235   The material that's used as fuel for making nuclear energy

wind gen er a tor   A machine that makes electricity from wind energy

Glossary
Electricity

ap pli ance   Something that uses electricity

bare   Not covered

bat ter y   Something that makes electricity

bulb   Something that lights up

cir cuit   A path for electricity

cir cuit break er   Something that stops the flow of electricity; a kind of fuse

con duct   To carry something, such as electricity

con duc tor   Something that carries electricity

con trol   To turn on or off

cord   Two covered wires that are stuck together

fil a ment   A thin wire inside a light bulb

flow   To move in one direction

fuse   Something that keeps a circuit safe by stopping the flow of electricity

fuse box   A place where the fuses for a house are found

glows   Gives off heat and light

in can des cent   Giving off light from a wire that glows

non con duc tor   Something that can't carry electricity

out let   A place to plug in something that uses electricity

o ver load ed   Carrying too much electricity

posts   The metal parts where wires are connected to a battery or a socket

shock   A sudden, sharp pain caused by electricity

short cir cuit   A short path for electricity

switch   Something that opens and closes a circuit

Glossary
Sound

---

ab sorb To soak up something, such as sound waves

a cous ti cal tile Special building material that has many tiny holes which absorb sound waves

a cous tics The science of sound control

am pli fy To make louder

ech o A sound that bounces back to your ears

fre quen cy How fast something vibrates

gas Something, such as air, that is not a solid or a liquid

liq uid Something that is wet, such as water

mo tion Movement

ob ject A thing

pitch How high or low a sound is

re flect To throw back something, such as sound waves

sol id Something that is hard, such as metal or wood

sound proof To keep sound from going in or out of a place

sound waves The way sound travels

sur face The outside of something

sur round ing All around

vi bra tion A back-and-forth movement

Glossary
Machines

con cen trate To put all of a force in one very small place

con trol To make something go as fast or slowly as you want

force A push or pull

fork lift A machine that's used to lift heavy things to very high places

ful crum The part a lever moves on

gear A kind of wheel and axle

groove The part of a pulley that a rope or belt fits in

hand truck A machine that's used to lift and move heavy things; it has two wheels

in clined plane A simple machine; it helps you move things from the ground to a high place

in crease To make more

lad der A machine that helps you climb to a high place

le ver A simple machine; it helps you move things that are tightly stuck; it also helps you lift heavy things

ma chine Something that helps you work

ob ject A thing

pul ley A simple machine; it helps you lift and move things

pul ley sys tem Two or more pulleys used together

ramp One kind of inclined plane

sim ple ma chine An inclined plane, wedge, lever, wheel and axle, and pulley

slope The way an inclined plane slants

split To push two sides of a thing apart. *You use an ax to split wood.*

steep Having a very high slant. *The sides of a mountain are steep.*

tool A small machine such as a hammer, saw, knife, and scissors

wedge A simple machine; it helps you cut, split, and make holes

wheel and ax le A simple machine; it helps you move things from one place to another

work To move something from one place to another

DATE DUE			
SEP 0 4			
SEP 10			
SEP 17			
OCT 25			
GAYLORD			PRINTED IN U.S.A.